ES   ..
.METEOROLOGY

THE WYKEHAM SCIENCE SERIES

General Editors:

PROFESSOR SIR NEVILL MOTT, F.R.S.
Emeritus Cavendish Professor of Physics
University of Cambridge

G. R. NOAKES
Formerly Senior Physics Master
Uppingham School

The Authors:

D. H. MCINTOSH is Senior Lecturer in the Meteorology
Department of the University of Edinburgh. In addition to
his teaching and research experience there he has supervised
many Field Studies courses attended by students of varied
academic background. He graduated in Physics at the
University of St. Andrews in 1939 and was a member of the
British Meteorological Office staff from then till 1955 during
which time his experience included weather forecasting in
different parts of the world and work at observatories.
He is author of 'The Meteorological Glossary'.

A. S. THOM graduated in Physics at the University of Glasgow
in 1964. After a year in Indiana as Teaching Assistant at
Rose Polytechnic Institute he came to the University of
Edinburgh where he obtained a Ph.D. degree in Meteorology
and is now Lecturer in that subject. Although his research
is primarily micrometeorological he has had since his early
years a great interest in weather phenomena in general and
enjoys communicating his enthusiasm to others.

ESSENTIALS OF
METEOROLOGY

D. H. McIntosh – University of Edinburgh

A. S. Thom – University of Edinburgh

TAYLOR & FRANCIS LTD
LONDON
1983

First published 1969 by Wykeham Publications (London) Ltd.

Reprinted 1972, 1973 (with minor corrections), 1978, 1981 and 1983.

Cover illustration—by courtesy of Keystone Press Agency Ltd.

ISBN 0 85109 040 0

Printed and bound in Great Britain by Taylor & Francis (Printers) Ltd. Rankine Road, Basingstoke, Hants.

PREFACE

THE readers for whom this book is primarily intended are students in the Sixth Form of schools and in an initial course at a university. Our main objective is to provide an understanding of meteorology as a physical science; further aims are to stimulate interest in making observations and measurements of atmospheric conditions and in relating local weather to the weather map.

In several chapters we have assumed that the standard of physics and mathematics attained by the reader is about that of the Sixth Form. In general, however, the demands made are less than this and most of the subject matter should be well within the compass of students of a general science background at upper school and early university levels.

Although the mathematics employed involves practically nothing more complicated than elementary differentiation and integration, the use of the partial derivative—written, for example, $\partial T/\partial t$—may not be familiar. In this example the derivative signifies, in fact, the rate of change of temperature (T) with time (t) when the other variables on which T depends (position coordinates) are held constant; $\partial T/\partial t$ is thus the *local* time rate change of T. The corresponding total derivative, written dT/dt, has a different physical meaning, namely the time rate of change of T of an individual (moving) mass of air. The relationship between such derivatives is explained in problem 8.4 and § 11.2.1. For the rest, it is sufficient to accept that, so far as the manipulations of differentiation and integration in this book are concerned, the two types of derivative are treated alike.

SI units are used throughout; symbols and units, and various numerical values, are set out on preliminary pages. The few departures from the SI system are in accordance with current meteorological practice and are as follows: (*a*) use of the millibar (rather than the $N\,m^{-2}$) as the unit of pressure; (*b*) use of the symbol R^*, rather than R, for the universal gas constant (R being used for the specific gas constant for dry air); (*c*) use of θ for potential temperature; (*d*) occasional use of T for temperature in °C when there is no risk of confusion (at all other times T stands for temperature in K); (*e*) use of the knot, i.e. the nautical mile per hour, for wind speed, in synoptic meteorology (for purposes of calculation $1\ m\ s^{-1} \approx 2$ knots).

Various chapters contain numerical examples, the answers to which are given at the end of the book.

Mr. V. T. Saunders, a schoolmaster of long experience, was consulted in the planning of the book; we much regret that he did not live to see its completion.   Mr. G. R. Noakes subsequently read the proofs and made valuable suggestions for which we are grateful.

We should also like to thank Mr. James Paton and Mr. R. W. Riddaway for reading and commenting on parts of the text.

Meteorology Department,   D. H. McIntosh
University of Edinburgh.   A. S. Thom
June 1969.

SYMBOLS, UNITS AND NUMERICAL VALUES

Quantity	Symbol	Magnitude	Unit
Constant of universal gravitation	G	$6 \cdot 67 \times 10^{-11}$	N m^2 kg^{-2}
Gas constant (per kmol)	R^*	$8 \cdot 314 \times 10^3$	J kmol^{-1} K^{-1}
Stefan's constant	σ	$5 \cdot 70 \times 10^{-8}$	W m^{-2} K^{-4}
von Kármán's constant	k	$0 \cdot 40$	
Latent heat of fusion of ice		$3 \cdot 35 \times 10^5$	J kg^{-1}
Latent heat of vaporization of water (15°C)	$L_v, (L)$	$2 \cdot 47 \times 10^6$	J kg^{-1}
Molecular weight of water vapour	M_v	$18 \cdot 0$	kg kmol^{-1}
Mean molecular weight of dry air	$\bar{M}, (M_d)$	$29 \cdot 0$	kg kmol^{-1}
Gas constant (per kg) for dry air	R	$2 \cdot 87 \times 10^2$	J kg^{-1} K^{-1}
Specific heat of air at constant pressure	c_p	$1 \cdot 01 \times 10^3$	J kg^{-1} K^{-1}
Standard gravitational acceleration	(g_0)	$9 \cdot 80665$	m s^{-2}
Standard atmospheric pressure	(p_0)	$1013 \cdot 25$	mb (1 mb $= 10^2$ N m^{-2})
Dry adiabatic lapse rate	Γ	$9 \cdot 8$	K km^{-1}
Saturated adiabatic lapse rate	γ_s		K km^{-1}
Earth's angular velocity	Ω	$7 \cdot 29 \times 10^{-5}$	(rad) s^{-1}
Mean Earth radius		$6 \cdot 367 \times 10^3$	km
Mean solar distance		$1 \cdot 49 \times 10^8$	km
Solar constant	S	$1 \cdot 39$	kW m^{-2}

Quantity	Symbol	Unit
Temperature	T	K, (°C)
Thermodynamic wet-bulb temperature	T_W	K, (°C)
Wet-bulb temperature	T_W'	K, (°C)
Dew-point temperature	T_d	K, (°C)

Quantity	Symbol	Unit
Virtual temperature	T_v	K, (°C)
Potential temperature	θ	K, (°C)
Wet-bulb potential temperature	θ_w	K, (°C)
Pressure	p	mb
Vapour pressure	e	mb
Mixing ratio	r	kg kg^{-1}, (g kg^{-1})
Relative humidity	U	(%)
Density	ρ	kg m^{-3}
Wavelength	λ	μm
Latitude	ϕ	
Coriolis parameter ($2\Omega \sin \phi$)	f	(rad) s^{-1}
Divergence (velocity)	div **V**	s^{-1}
Vorticity (vertical)	ζ	s^{-1}
Shearing stress	τ	N m^{-2}
Mixing length	l	m
Friction velocity	u_*	m s^{-1}
Roughness length	z_0	mm
Transfer coefficient	C	
Richardson's number	Ri	

CONTENTS

Preface   v

Symbols, units and numerical values   vii

Chapter 1 THE NATURE AND SCOPE OF METEOROLOGY

1.1. Meteorology in relation to other sciences   1
1.2. Variations in space and time   3
1.3. Applied meteorology   4

Chapter 2 PHYSICAL PROPERTIES OF THE ATMOSPHERE

2.1. Composition of dry air   5
  2.1.1. Mean molecular weight   6
  2.1.2. Dissociation and ionization   7
  2.1.3. Escape to space of component molecules   7

2.2. Pressure, density and temperature   9
  2.2.1. Definition of pressure   9
  2.2.2. Values near sea level   9
  2.2.3. Variations in the vertical   9
  2.2.4. Diurnal fluctuations at upper levels   12
  2.2.5. Horizontal pressure gradients   14

2.3. Water vapour   15
  2.3.1. Humidity mixing ratio   16
  2.3.2. Density of moist air   17
  2.3.3. Saturation vapour pressure   17
  2.3.4. Paths leading to saturation   18
  2.3.5. Measurement of vapour pressure   20
  2.3.6. Distribution of water vapour   20

Chapter 3 HEAT TRANSFER

3.1. Radiation processes   23
  3.1.1. Solar radiation: its energy distribution   23
  3.1.2. The solar constant   24
  3.1.3. Effect of the atmosphere and Earth on solar radiation   25
  3.1.4. Radiation from the Earth and atmosphere   28

3.2. Convection	32
3.2.1. Adiabatic temperature changes	33
3.2.2. Adiabatic equation	33
3.2.3. Potential temperature: dry adiabatic lapse rate	35
3.2.4. Saturated adiabatic lapse rate	36
3.2.5. Stability and instability	36
3.3. Heat transfer in land and sea	38
3.3.1. Heating and cooling of soil	39
3.3.2. Heating and cooling of water	39

Chapter 4   CONDENSATION AND PRECIPITATION

4.1. Microphysical processes	41
4.1.1. Condensation nuclei	41
4.1.2. Curvature and solute effects	42
4.1.3. Water-droplet clouds	42
4.1.4. Ice nuclei	43
4.1.5. Ice-crystal clouds	43
4.1.6. Precipitation from water clouds	43
4.1.7. Precipitation from mixed clouds	44
4.1.8. Thunderstorm electricity	45
4.2. Larger-scale processes	46
4.2.1. Surface cooling	46
4.2.2. Evaporation	48
4.2.3. Vertical motion	49
4.3. Cloud observations	54
4.3.1. Cloud genera: their heights and composition	55
4.3.2. Cloud recognition and general features	55
4.3.3. Effects of vertical wind shear	56
4.3.4. Cloud classification for forecasting	57

Chapter 5   THE TEPHIGRAM

5.1. Construction of the diagram	60
5.1.1. Coordinates: area and energy	60
5.1.2. Isobars	61
5.1.3. Saturation mixing ratio lines	61
5.1.4. Saturated adiabatics	62
5.1.5. Height variation	64

5.2. Simple graphical computations 64
  5.2.1. Height 64
  5.2.2. Humidity elements 65
  5.2.3. Condensation levels 67
  5.2.4. Föhn effect 68

5.3. Precipitable water and precipitation rate 68
  5.3.1. Formula and calculation 68
  5.3.2. Precipitation rate 69
  5.3.3. Water content of convection clouds 69

5.4. The effects of vertical motion on lapse rate 70
  5.4.1. Unsaturated or saturated motion 70
  5.4.2. Potential (convective) instability 71

5.5. Tephigram analysis 72
  5.5.1. Latent instability 72
  5.5.2. Air mass characteristics 74

Chapter 6 WINDS
6.1. Laws of motion and the Earth's rotation 78
  6.1.1. Newton's First and Second Laws 78
  6.1.2. Nature of the Earth's rotation 79
  6.1.3. Effects of the Earth's rotation: the Coriolis force 80

6.2. Inertial flow and geostrophic winds 83
  6.2.1. Nature of inertial flow 83
  6.2.2. Nature of geostrophic flow 83
  6.2.3. Geostrophic wind equation 84
  6.2.4. Wind and pressure near the equator 87

6.3. Gradient winds 88

6.4. Winds in the friction layer 90

6.5. Thermal winds 91
  6.5.1. Vertical shear vector 91
  6.5.2. Temperature control of the shear vector 91
  6.5.3. Thermal wind equation and thickness charts 93
  6.5.4. Hodographs and temperature advection 94
  6.5.5. Jet streams 96

Chapter 7 INSTRUMENTS AND OBSERVATIONS

7.1. Routine surface observations	99
7.1.1. Pressure	99
7.1.2. Temperature and humidity	100
7.1.3. Precipitation and evaporation	102
7.1.4. Wind	103
7.1.5. Clouds and visibility	104
7.1.6. Sunshine and radiation	104
7.1.7. Ship observations	105
7.2. Upper air observations	105
7.2.1. Historical	105
7.2.2. The radiosonde: radar winds	109
7.2.3. Ozone measurements	111
7.3. World Weather Watch	112
7.4. Experiments in observation and interpretation	114
7.4.1. Pressure	114
7.4.2. Temperature and humidity	114
7.4.3. Evaporation and rainfall	115
7.4.4. Wind	116
7.4.5. Radiation	117
7.4.6. Topographical influences	118

Chapter 8 SYNOPTIC METEOROLOGY

8.1. The surface weather map: an introduction	119
8.1.1. The plotting code	119
8.1.2. Pressure systems and features	122
8.1.3. Air masses	123
8.1.4. Fronts	124
8.2. Air mass characteristics	126
8.2.1. Classification	126
8.2.2. Modifications	127
8.2.3. Air masses over the British Isles	127
8.3. Frontal characteristics	129
8.3.1. The stability of a frontal surface	129
8.3.2. Equilibrium slope of a frontal surface	130
8.3.3. Frontal structure	132

xii

8.4. Frontal depressions   134
   8.4.1. The life cycle of a frontal depression   135
   8.4.2. Cold front waves; depression families   137
   8.4.3. Warm front waves   137
   8.4.4. Secondaries at points of occlusion   139

8.5. Non-frontal depressions   139
   8.5.1. Heat lows   139
   8.5.2. Polar lows   140
   8.5.3. Orographic lows   141
   8.5.4. Tropical cyclones   142
   8.5.5. Tornadoes   143

8.6. Anticyclones   143
   8.6.1. General characteristics   143
   8.6.2. Cold and warm anticyclones   144

8.7. Synoptic development   145
   8.7.1. Convergence, divergence and vertical motion   145
   8.7.2. Convergence and vorticity   149
   8.7.3. Long waves   154
   8.7.4. Circulation indices: blocking   155

8.8. Surface analysis   156
   8.8.1. General   156
   8.8.2. Representativeness of observations   158
   8.8.3. METMAPS   159

Chapter 9 MICROMETEOROLOGY
9.1. The nature of airflow near the ground   164
   9.1.1. Wind speeds over a uniform level surface   164
   9.1.2. Flow within a fluid boundary layer   165
   9.1.3. Shearing stress via the mixing length concept   166
   9.1.4. The friction velocity u_* 167
   9.1.5. Interpretation of the mixing length concept   167
   9.1.6. The wind profile equation in complete form   168

9.2. The influence of surface roughness on the wind   168
  9.2.1. Roughness in the aerodynamic sense   168
  9.2.2. Roughness in relation to shearing stress and mean wind speed   169
  9.2.3. The drag coefficient C_D 171
  9.2.4. C_D as a transfer coefficient   171
  9.2.5. Effect of a change in surface roughness   172

9.3. Vertical transport by turbulence   173
  9.3.1. Flux equations; use of electrical analogy   174
  9.3.2. Heat flux and other calculations   176
  9.3.3. Vertical temperature gradients in relation to turbulent exchange   178

Chapter 10 THE GENERAL CIRCULATION
10.1. General characteristics   184
  10.1.1. Genesis and interactions   184
  10.1.2. Time fluctuations   185

10.2. Observations   185
  10.2.1. Time- and space-averaging   185
  10.2.2. Tracers   190

10.3. Experiment and theory   195
  10.3.1. The rotating vessel experiment   195
  10.3.2. Conservation principles   196
  10.3.3. Cellular models   198

10.4. Climatic zones   201

Chapter 11 WEATHER FORECASTING
11.1. Historical survey   203
  11.1.1. 1860–1920   203
  11.1.2. 1920–1945   203
  11.1.3. 1945–1960   204
  11.1.4. 1960 onwards   205

11.2. Conventional forecasting   208
  11.2.1. Pressure tendency   208
  11.2.2. Making the forecast   210

11.3. Long-range forecasting 213
    11.3.1. Statistical methods 214
    11.3.2. Synoptic methods 215
    11.3.3. Analogues 216

11.4. Numerical forecasting 217
    11.4.1. The barotropic model 217
    11.4.2. Later developments 220

11.5. Predictability and control 220
    11.5.1. Short-range predictability 220
    11.5.2. Medium-range predictability 221
    11.5.3. Long-range predictability: climatic trends 222
    11.5.4. Weather and climate modification 224

Answers to Problems 227

Subject Index 229

The Wykeham Series 239

CHAPTER 1
the nature and scope of meteorology

1.1. *Meteorology in relation to other sciences*
METEOROLOGY is a branch of physical science in which a fundamentally different approach is required from that adopted by the laboratory physicist. The latter's method of solving problems is to ' divide and rule '; that is to say, he solves a particular problem by first breaking it down into its component parts and then, by making experiments under conditions of strict control, he determines the rule which governs each separate part: consider, for example, the thermodynamic laws which connect the pressure, volume and temperature of a gas, or the dynamical laws which relate force, mass and acceleration.

The meteorologist's approach is, of necessity, very different. The atmosphere is his laboratory, but the ' experiments ' which are continuously enacted there are—doubtless fortunately!—beyond his control, He has no way of isolating the various processes; he cannot, for example, bring a mass of air to a halt, shelter it from solar radiation and prevent water from evaporating into it, while he proceeds to measure the heat exchanges between the mass and its immediate surroundings. He must, on the contrary, adopt a comprehensive view of all the processes involved, while at the same time distinguishing and if possible evaluating the individual processes that comprise the whole. In this latter respect his task is made immensely more difficult by the fact that the various effects do not add and subtract in a simple way, i.e. they are not independent but, on the contrary, are linked each with all the others. The meteorologist's consolation is that he is at least able to accept the laws of classical physics and is not required to verify them in the atmosphere!

Mathematics plays very much the same role in meteorology as it does in physics; that is to say it provides a very convenient and powerful method of short-hand expression of physical ideas and provides also the means by which these ideas are given numerical form (although, in meteorology, the lack of independence between the various processes introduces special difficulties in this latter respect). It is true, nevertheless, that—as in physics—most atmospheric processes can be comprehended in descriptive terms and without recourse to mathematics.

Meteorology has relatively little in common with those of the Earth sciences which are concerned with conditions within the Earth itself, namely geology, seismology and geomagnetism.† On the other hand, its connection with oceanography is extremely close because the atmosphere to a large extent drives the oceans and, paradoxically, receives a large part of the energy which enables it to do so from the water that is evaporated from the oceans. Hydrology, which is concerned with water in all its forms from the time it is precipitated upon the land until it is discharged into the sea or is returned to the atmosphere, is another science with which meteorology is closely associated.

If the oceans are to be regarded as the atmosphere's main boiler-house then the highly-conducting ionosphere, which extends upwards from about 60 km, may be regarded as its electrical store-house. The air which is present at these levels is extremely tenuous and, since many of the particles are charged, the motion there is governed less by pressure forces than by the earth's magnetic field. So far as weather—as it is generally understood—is concerned, such high regions of the atmosphere are of academic rather than of immediate practical interest. However, if we adopt the wider definition of meteorology as the study of the phenomena of atmospheres, then our interest extends not only to the whole of the Earth's atmosphere, but also to those of the planets and even to that of the sun itself. Thus the meteorologist who is concerned with a broad view of the atmosphere's functioning, or with detailed phenomena in the Earth's high atmosphere, has a good deal of common ground—so to speak—with the astronomer and with the space scientist.

Since the atmosphere is a mixture of gases that is heated and cooled and is in continuous motion, the branches of physics which are most directly concerned in its functioning are thermodynamics and fluid dynamics. Other physical topics are involved in varying degree; also chemistry, for example, in relation to the composition of clouds and precipitation, to atmospheric pollution and to the photochemical processes that occur in the high atmosphere. Furthermore, meteorological data accumulate at such a rate that their ' digestion ' and treatment require the application of the most modern methods of data processing and numerical analysis. The complete meteorologist, if such exists, would obviously have to be a man of many parts.

Conversely, meteorological processes have to be taken into account in many sciences other than those already mentioned. Thus, one of the atmospheric gases, carbon dioxide, is taken up by plants in the presence of sunlight (photosynthesis) to form their essential tissue; the carbon dioxide reaches the plant by turbulent diffusion downwards through a fluid boundary layer while the same turbulence is responsible for the

† There are, however, exceptions such as the parts played by mountain building and by the (controversial) drift of continents in relation to past climates; also the small, rapid variations of the earth's magnetic field, measured at the ground, which are, in fact, produced by motions in the earth's electrically-conducting high atmosphere.

transport of airborne microbes. Such processes are therefore of concern in soil physics, plant physiology and microbiology. Many other examples might be quoted.

1.2. *Variations in space and time*

We are well aware that the weather we experience at any given time is not a completely reliable guide as to what may quite shortly transpire; equally, that it may not be representative of what lies just beyond our immediate horizon. These facts emphasize the complexity, in space and time, of the system with which the meteorologist deals; he can describe the state of the system—present as well as future—only to a greater or lesser degree of approximation.

Some order may be brought into the apparent chaos, which is not an inappropriate description of atmospheric motion as a whole, by recognition of a somewhat natural division of scale. Smallest of all is the molecular scale on which, in the ultimate, all atmospheric exchanges occur. This scale does not enter directly into the meteorologist's calculations, or even his thinking; he is concerned always with the larger, bulk properties of the atmosphere. Smallest of these is the ' micro ' scale in which elements of air upwards from 1 mm or so in cross-section are the important agents.

Beyond this scale of micrometeorology, which we consider in Chapter 9, and the familiar ' synoptic ' scale, which is considered mainly in Chapter 8 but also elsewhere, is an intermediate or ' meso ' scale. This includes such phenomena as tornadoes, thunderstorms, and local cloud formations and wind circulations. A good deal is known about these phenomena in general but, by their very nature, they are mainly unrecorded so far as ' current ' weather is concerned, information about this being provided by a network of reporting stations which, even overland, are rarely closer than about 50 km.

On the synoptic scale we have the familiar depressions and anticyclones, varying greatly in their dimensions and life spans. If time averages are considered, then the less persistent systems are removed while the more persistent features emerge clearly and represent a ' general circulation ' of the atmosphere, discussed in Chapter 10.

Such a separation of atmospheric motion into micro, meso, synoptic and general categories is very convenient for practical purposes. This is not, however, to say that these categories are in any real sense distinct from each other; on the contrary each scale influences, and in turn depends on, all the others. This is a point which was elegantly made by L. F. Richardson in the following terms:

> ' Big whirls have little whirls
> Which feed on their velocity.
> Little whirls have lesser whirls
> And so on to viscosity '.

1.3. *Applied meteorology*

Since atmospheric conditions impinge in one way or another on almost every human activity, direct benefit is gained from the application of knowledge which relates to these conditions. This may take the form of the use of climatological data or the application of general principles; in many cases, however, prediction is involved.

According to popular belief, meteorology is weather forecasting and all meteorologists are weather forecasters. Neither view is strictly correct but there is some justification for them in that—certainly from the point of view of economics—prediction is the most important activity on which professional meteorologists are engaged; also in the fact that most meteorological research is aimed, directly or indirectly, at improving the accuracy of weather forecasts and extending the period for which they are applicable.

It has often been pointed out that meteorologists are placed in an awkward and unusual position, for scientists, in that they are expected to make predictions concerning a physical system over which they exert no control and of which they have imperfect understanding, not to mention information. This situation arises because—despite their well-known shortcomings—weather forecasts are economically extremely valuable; for example their estimated annual value to agriculture in the United Kingdom is of the order of £20 million and similar figures apply to many other fields of activity. Correspondingly greater benefits would result from the improvements towards which meteorological research is aimed.

CHAPTER 2
physical properties of the atmosphere

A SUBJECTIVE conception of the atmosphere is that of a generally transparent medium endowed with ' temperature ', capable of motion and liable to produce rain. However, to predict occurrences of fog, ' cold ', wind, rain or any one or combination of a multitude of weather phenomena, the meteorologist must be in the first instance objective: his conception of the atmosphere must be based on a firm understanding of its physical properties. He must know values of pressure, density and temperature at each of many points both in the vertical and the horizontal and be able to relate those values to the present physical state of the atmosphere, thence to predict its future behaviour. He must know and understand the significance of the gaseous composition of the atmosphere, recognizing especially the importance of water vapour.

2.1. *Composition of dry air*

The composition of the atmosphere is remarkably uniform. If an air sample taken from any region within 80 km of the Earth's surface is analysed, the relative proportions of its major constituent gases are found to vary by no more than a few thousandths of one per cent—provided that any water vapour is first removed. Such close uniformity in composition implies that up to 80 km or so there must be sufficient large-scale vertical mixing to counter the natural tendency of the constituent gases to separate out according to molecular weight (diffusive separation). Above 80 km such mixing, if not altogether absent, is certainly weak enough to allow molecular diffusion to begin to effect a redistribution of the individual gases, there being progressively greater proportions of the lighter constituents with increase in height.

Thus most of the air in our atmosphere—over 99·999% by weight of which lies below 80 km—is a thorough mixture of gases. The most abundant constituent is nitrogen (N_2), followed by oxygen (O_2), argon (A) and carbon dioxide (CO_2). These gases are present in the percentage ratios 78·09, 20·95, 0·93 and 0·03, by volume, respectively. There are also measurable traces of other gases, such as ozone (O_3) which has effects far outweighing its meagre proportional representation (see § 3.1).

5

2.1.1. *Mean molecular weight*

If $\bar{M}$ is a suitably defined mean value of the molecular weights of the individual gases in a mixture then the pressure p, temperature T and density ρ of the mixture can be related by the gas equation in the form:

$$p = \rho \cdot \frac{R^*}{\bar{M}} \cdot T, \qquad (2.1)$$

where R^* is the universal gas constant, equal to $8 \cdot 314 \times 10^3$ J kmol^{-1} K^{-1}. The quantity $\bar{M}$ is not the simple weighted average value given, for a mixture of two gases, by the expression $(m_1 M_1 + m_2 M_2)/(m_1 + m_2)$ in which M_1 and M_2 are the respective molecular weights of the gases and m_1 and m_2 the masses of each present, but may be derived as follows. Let the mixture be contained in a vessel of volume V. The partial pressure of the first gas is given by

$$p_1 = \frac{m_1}{V} \cdot \frac{R^*}{M_1} \cdot T, \qquad (2.2)$$

m_1/V being its contribution to the total density of the mixture ρ, while

$$p_2 = \frac{m_2}{V} \cdot \frac{R^*}{M_2} \cdot T. \qquad (2.3)$$

As $p = p_1 + p_2$ and $\rho = (m_1 + m_2)/V$ we see from equations (2.1), (2.2) and (2.3) that for a mixture of any number of gases

$$\bar{M} = \frac{m_1 + m_2 + \cdots}{\dfrac{m_1}{M_1} + \dfrac{m_2}{M_2} + \cdots}; \qquad (2.4)$$

i.e. $\bar{M}$ is defined as the weighted harmonic mean of the molecular weights of the gases present.

Throughout the lowest 80 km of the atmosphere $\bar{M} = 28 \cdot 966$ kg kmol^{-1}, while above 80 km $\bar{M}$ is known to decrease progressively—but at exactly what rate is not yet known.

At the highest levels neither $\bar{M}$ nor T is separately determined as yet to any high degree of accuracy, although the ratio $T/\bar{M}$ is known within reasonable limits from independent measurements of pressure and of density (see equation (2.1)). The minute air densities which exist at such great heights augment the usual difficulty encountered in measuring air temperature—namely that the temperature of a body immersed in a fluid will differ from that of the fluid if it (the body) is gaining or losing heat by exchange of radiation with its surroundings. This temperature difference depends on the efficiency with which the fluid itself can exchange heat with the body, by way of conduction and convection, processes which depend directly on the density of the fluid. At great heights there are just not enough air molecules to ensure that the

6

temperature of the air is even approximately communicated to a sensing element, whose temperature must therefore depend on exchange of radiation with its surroundings. Thus temperature in the upper atmosphere bears little relationship to that of an immersed body and must be looked upon merely as defining the mean kinetic energy of the molecules present.

2.1.2. Dissociation and ionization

It does appear however that $\bar{M}$ tends towards a value, at the highest levels in the atmosphere, close to one-half of its value in the lowest 80 km. This suggests that in the upper atmosphere both nitrogen and oxygen are present largely in atomic form. However, although oxygen is readily dissociated by the action of ultra-violet radiation from the sun, nitrogen is not. (Above 200 km oxygen is almost completely dissociated, while even at 400 km little nitrogen is in atomic form.) Part of the observed decrease in mean molecular weight must therefore be due to diffusive separation of N_2, N and O and also, above about 800 km, to the increased presence similarly brought about, of helium (He) and of hydrogen (H).

Absorption of solar ultra-violet radiation by molecules in the upper atmosphere may result in ionization rather than dissociation, so that there is an increasing relative abundance of positively charged ions and of electrons upwards of 60 km. The actual density of electrons however reaches a well-marked maximum at about 300 km, although a number of indistinct maxima occur at intermediate levels. The region between 60 and 300 km, known as the ionosphere, has properties critical to the propagation of radio waves. Long-distance radio communication is made possible by the reflection of radio waves at discontinuities in the gradient of the electron density curve: however, during periods of abnormally high (local) values of electron density, the electrons themselves absorb so much energy from the waves as to cause radio fade-out. The positively charged ions, being many times heavier than the electrons and correspondingly less mobile, contribute little to the peculiar properties of the ionosphere: they do however represent an intermediate stage in several chain reactions leading independently to dissociation, e.g. recombination of a positive ion of molecular oxygen (O_2^+) with an electron (e^-) leads to two neutral oxygen atoms ($O+O$).

2.1.3. Escape to space of component molecules

We are now in a position to appreciate why our atmosphere contains considerably more nitrogen than oxygen and why helium and hydrogen are so poorly represented therein. On the fringe of the atmosphere, air can no longer properly be termed a gas; individual particles perform long elliptical orbits between collisions. However, if a given collision is violent enough to impart to one of the participants a velocity greater

7

than or equal to the escape velocity v_e (at that level) and if this imparted velocity is in any sense upwards, then, barring further collisions, the particle will be projected out of the atmosphere. For this reason the term exosphere is given to the atmospheric fringe region (above 600 km). It has been shown that if the root mean square velocity c of the molecules of a gas in the exosphere is a small enough fraction of v_e —say 0·1—its rate of loss will be entirely negligible, but that if the fraction c/v_e increases much beyond 0·2 the rate of loss rapidly becomes significant.

Table 2.1 gives values for the planetary half-lives ($t_{1/2}$) of the exospheric constituents, derived from values calculated by Spitzer but using the presently accepted temperature $T_e = 1500$ K for the exosphere. ($t_{1/2}$ depends strongly on temperature, the mean square velocity of a gas being directly proportional to its temperature.) Thus half the hydrogen in the atmosphere is lost† every 10 000 years—during but a scrap of geological time—while nitrogen's unwillingness to dissociate to any appreciable extent ensures permanence or even enhancement of its present position as the most abundant constituent of our atmosphere.

	M	c/v_e	Earth $t_\frac{1}{2}$ years
H	1	0·57	10^4
He	4	0·28	10^{10}
N, O	14, 16	0·15	$\sim 10^{35}$
N$_2$	28	0·11	$\sim 10^{60}$
		For $T_e = 1500$ K	

Table 2.1. Planetary half-lives ($t_\frac{1}{2}$) of exospheric constituents.

Finally, the rate at which helium is produced by decay of radioactive elements in the Earth's crust is known. Thus, provided the total amount of helium in the atmosphere has varied little on a geological time scale, this rate of production may be equated to the rate at which helium is being lost from the atmosphere, and a value of $t_{1/2}$ deduced which is in order-of-magnitude agreement with that quoted in Table 2.1. An indirect confirmation of the mean temperature in the exosphere may thus be obtained, within limits of about ± 200 K. This method was in fact used to make earlier estimates of kinetic temperature in the outermost regions of our atmosphere.

† Assuming equilibrium much the same amount must be gained, presumably by interception of protons (H^+) from the solar atmosphere or by decomposition of water vapour in the upper atmosphere by solar ultra-violet radiation.

2.2. Pressure, density and temperature

2.2.1. Definition of pressure

In the preceding section pressure is introduced as a formal component in the gas law equation (equation (2.1)). As such, it represents the force acting normally on unit area of a body immersed in a gas, or mixture of gases—a force resulting from the 'bouncing' of countless thermally-excited molecules against the body. So defined, pressure at a point in a gas must be the same in all directions. This must not be forgotten in the face of the practical meteorologist's definition; namely that pressure (p) at a point in the atmosphere is simply the weight of air vertically above unit horizontal area centered on the point—i.e. the downward force on unit horizontal area resulting from the action of gravity (g) on the mass (m) of air vertically above.

2.2.2. Values near sea level

Atmospheric pressure has an average sea-level value close to 1 bar or 10^5 N m^{-2} which corresponds, with $g = 9 \cdot 81$ m s^{-2}, to a mass $m = p/g = 1 \cdot 02 \times 10^4$ kg of air (or about 10 tons) above every square metre of the Earth's surface.

A representative value for air density close to sea level may be found from equation (2.1) in the form

$$\rho = p/RT, \tag{2.5}$$

in which R is the specific gas constant for dry air given by $R^* \div \bar{M}$. Equating p to 10^5 N m^{-2} and T to 288 K (15°C), its average sea-level value, we find that $\rho = 1 \cdot 2$ kg m^{-3}. It follows that if air were incompressible the depth (h) of the atmosphere would be limited to $m \div \rho = 8 \cdot 4$ km. In another vein, this would be the height of a hypothetical 'incompressible-air' barometer (assuming negligible variation of g in the vertical).

However, the real atmosphere is not incompressible, so that ρ as well as p must decrease upwards, indefinitely, and h is therefore indeterminate. Also, most real barometers contain mercury, whose density exceeds that for air, derived above, by a factor close to $1 \cdot 1 \times 10^4$. Thus total atmospheric pressure may be balanced against a mercury column only $(8 \cdot 4 \times 10^3 \text{ m}) \div 1 \cdot 1 \times 10^4 = 0 \cdot 76$ m in height (or about 30 in). Correspondingly, this would be the depth of a hypothetical mercury 'atmosphere' at whose base $p = 1$ bar.

2.2.3. Variations in the vertical

The relationship tacitly assumed in the preceding argument is that

$$p = g\rho h. \tag{2.6}$$

This is known as the barometer equation and may be used to calculate the pressure at any depth h within an incompressible fluid. Equation

9

(2.6) is the integral of the hydrostatic equation

$$dp = -g\rho dz, \qquad (2.7)$$

obtained if ρ is held constant. (The minus sign ensures a pressure drop across the height increment dz.)

If, however, ρ is obtained for a compressible atmosphere from Equation (2.1), we may write

$$\frac{dp}{p} = \frac{-g\bar{M}}{R^*T} \cdot dz \qquad (2.8)$$

or equivalently, from equation (2.5),

$$\frac{dp}{p} = \frac{-g\,dz}{RT}. \qquad (2.9)$$

Equation (2.8) shows that the fractional rate of decrease of pressure with height,

$$\frac{1}{p}\frac{dp}{dz}$$

is directly proportional to $\bar{M}$, and inversely proportional to absolute temperature T. In the absence of large-scale vertical mixing the partial pressure of each constituent gas would decrease according to equation (2.8) at a separate rate, proportional to its own molecular weight. Under these conditions lighter gases would be present in greater ratio at higher levels and heavier gases at lower (so-called diffusive equilibrium) and $\bar{M}$ would decrease with z.

If, however, we assume that $\bar{M}$ is independent of z we may integrate equation (2.9) to give

$$p(z) = p_0 \exp(-gz/R\bar{T}), \qquad (2.10)$$

in which p_0 is surface pressure and $\bar{T}$ the mean temperature of the air up to the level z. As an estimate of $\bar{T}$ enables z to be determined from a knowledge of $p(z)$ and p_0, equation (2.10) is often referred to as the altimeter equation. Similarly equation (2.10) can be used to calculate the thickness ΔZ of a layer of air lying between two pressure levels— an 'isobaric layer'. Thus, if Z_1 and Z_2 are the heights of the pressure levels p_1 and p_2,

$$\Delta Z = Z_2 - Z_1 = \frac{R\bar{T}}{g} \cdot \ln(p_1/p_2), \qquad (2.11)$$

although the finite difference form of equation (2.9), namely

$$\Delta Z = \frac{R\bar{T}}{g} \cdot \frac{\Delta p}{\bar{p}} \qquad (2.12)$$

is easier to use and introduces little error provided p_1/p_2 is close to unity: $\bar{p}$ is the mean of p_1 and p_2, and Δp their difference. In both these equations $\bar{T}$ is the mean temperature (in K) of the layer, so that the thickness of a given isobaric layer is a direct measure of its mean absolute temperature. The concept of thickness is used in Chapter 6 to derive the so-called thermal wind component of the upper winds.

Tabulated average values of pressure and density are given in fig. 2.1 at several levels up to 100 km, together with a curve showing the corresponding variation in temperature appropriate to average mid-latitude conditions.

The various vertical sub-regions derive quite simply from the way in which temperature changes locally with height. Thus in the troposphere and mesosphere temperature (in general) decreases with

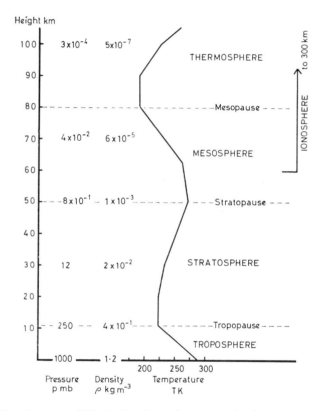

Fig. 2.1. Average mid-latitude values of pressure, density and temperature.

height; while in the stratosphere and again in the thermosphere there is a general increase in temperature upwards. Where temperature decreases with height there is said to be a positive lapse rate (lapse conditions): if temperature is constant in the vertical, conditions are said to be isothermal: while if temperature increases with height (a negative lapse rate) there is said to be an inversion.

11

In the troposphere, which contains all phenomena recognizable as weather, there is an average lapse rate of 6·5 K km^{-1}†, although lapse conditions need not exist throughout its entire depth at any given time or place. The boundary between the troposphere and stratosphere is known as the tropopause, while the terms stratopause and mesopause are applied to the corresponding boundaries above (see fig. 2.1).

The height of the tropopause is a function of latitude and season—as demonstrated by fig. 2.2. Day-by-day variations also occur, being largest and most frequent in mid-latitudes during winter. Note that the lowest temperatures in the troposphere are to be found not over the winter pole but over the equator, where the extreme height of the tropopause more than cancels the effect of high surface temperatures.

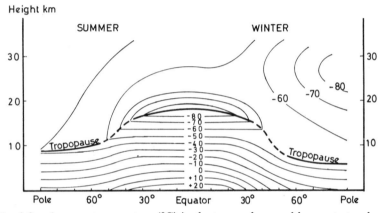

Fig. 2.2. Average temperature (°C) in the troposphere and lower stratosphere.

Other less explicable anomalies are found in the seasonal and latitudinal variations in temperature at certain levels above the troposphere. For example, although the middle and upper stratosphere is much warmer in summer than in winter the mesosphere seems to behave in exactly the opposite sense, its seasonal temperature minimum occurring in summer. It is likely, however, that this and other apparent anomalies can be explained by invoking the existence of (so far undetectable) vertical motions in the stratosphere and mesosphere.

2.2.4. *Diurnal fluctuations at upper levels*

Values of pressure, density and temperature at levels in excess of 100 km remain somewhat speculative—especially above 200 km where large diurnal fluctuations occur. For example, at 200 km, where the average temperature is known to be somewhere near 1200 K, there is a

---

† An exception to this occurs near the winter pole, where a strong surface inversion extends to several kilometres.

diurnal range in temperature of about 250 K and in density of about 5%; while at 600 km, where the thermosphere merges into the exosphere and temperatures average around 1500 K, the corresponding ranges are about 500 K and 600% respectively. The higher densities occur in phase with the higher temperatures and vice-versa, a phenomenon not entirely obvious until one realizes that large pressure rises must occur at upper levels in response to daytime heating, as follows.

Let z_1 and z_2 be two levels in the atmosphere, z_1 being the lower. Allow the mean absolute temperature of this layer to rise from T_1 to T_2, ensuring that the pressure at the lower level remains unchanged. It then follows from equation (2.10) that the pressure at the upper level will increase by the factor

$$\exp\left[\frac{g(z_2 - z_1)}{R T_1}\left\{1 - \frac{T_1}{T_2}\right\}\right] = F, \text{ say,}$$

i.e. by the percentage f, where $f = (F-1) \times 100\%$. For a given percentage rise in T the magnitude of F is strongly dependent on the depth $(z_2 - z_1)$ of the layer, relative to the size of the product RT/g known as the scale height H. (It is easily deduced from equation (2.10) that H is the increment in height above any level in the real atmosphere associated with a reduction in pressure to the fraction $1/e$ of its initial value; i.e. $p(z+H) = p(z)/e$. Also, substituting for the product RT from equation (2.5) one finds that $H = p/\rho g$ or m/g (see § 2.2.2); so that H may be looked upon as the thickness of the hypothetical layer of uniform density $\rho(z)$ with which the real atmosphere above the level z may be replaced, $p(z)$ being unaffected.) Table 2.2 gives the percentage rise in pressure—at the top of each of several layers of different depths—found by assuming a 25% increase in mean absolute temperature and an average scale height of 50 km, these being plausible mean values for the region under discussion, namely 100 to 600 km.

Layer depth $(z_2 - z_1)$ km	5	10	20	50	100	200	500
$(z_2 - z_1)/H$	0·1	0·2	0·4	1·0	2·0	4·0	10·0
Pressure change (f) %	2	4	10	25	57	146	850
Density change %	−18	−17	−12	0	+26	+97	+660

Table 2.2. Percentage changes in pressure and density at the top of an atmospheric layer subject to a 25% increase in mean absolute temperature. An average scale height $H = 50$ km is assumed.

The corresponding percentage changes in density, also included in Table 2.2, show that (uniform) heating of an atmospheric layer must produce a positive fluctuation in density as well as pressure at levels in excess of the scale height H above its base. In the upper regions of a

13

layer whose depth is more than a few scale heights the diurnal fluctuations in pressure and density are relatively very much larger than the diurnal temperature fluctuations which they accompany.

2.2.5. *Horizontal pressure gradients*

To the practising meteorologist, variation of pressure in the horizontal, although several orders of magnitude less than in the vertical, is the more important. For, unlike the large pressure gradient ' built in ' to the vertical by gravity and hence (conversely) balanced by gravity, any horizontal gradient of pressure represents a state of imbalance which must give rise to horizontal air motion, i.e. wind. Near sea level the magnitude of the vertical pressure gradient ($\partial p/\partial z$) is roughly 10 mb per 100 metres, whereas gradients of 10 mb per 100 *kilo*metres in the horizontal produce winds in excess of hurricane force.

If n is the horizontal direction in which pressure increases most rapidly, the horizontal pressure gradient is defined by $\partial p/\partial n$. Now, the units of pressure gradient are force per unit volume, so that $\partial p/\partial n$ is recognized as the force on unit volume of air required to produce and maintain the pressure gradient†. On the other hand, the force on the unit volume due to the existence of the pressure gradient is $-\partial p/\partial n$. This force is usually expressed per unit mass and termed the pressure gradient force (p.g.f.); thus

$$\text{p.g.f.} = -\frac{1}{\rho}\frac{\partial p}{\partial n}. \tag{2.13}$$

Horizontal pressure gradients in the atmosphere are produced indirectly, e.g. by continuing local divergence of air lying above. They are able, however, to maintain themselves in a state of dynamic, as distinct from static, equilibrium in which the p.g.f. is balanced by friction against the Earth's surface‡.

If the p.g.f. is not expressed directly in terms of the horizontal pressure gradient, but is related to the slope $\partial Z/\partial n$ of the corresponding isobaric surface, then an expression is obtained in which density does not appear explicitly, viz. from fig. 2.3:

$$\frac{\partial p}{\partial n} \doteq \frac{\Delta p}{\Delta n} = \frac{p_B - p_A}{\Delta n} = \frac{p_B - p_C}{\Delta n} = \frac{-(p_C - p_B)}{\Delta n}$$

$$= \frac{-\left(\dfrac{\Delta p}{\Delta Z}\right)\Delta Z}{\Delta n} \doteq -\frac{\partial p}{\partial z}\cdot\frac{\partial Z}{\partial n} = g\rho\frac{\partial Z}{\partial n}$$

from equation (2.7).

† In the vertical $n = -z$. Thus $\partial p/\partial n = \rho g$, which merely states that the force of gravity is responsible for the production and maintenance of the vertical pressure gradient.

‡ Properly it is the p.g.f. in the direction of motion which is balanced by friction, while the p.g.f. at right angles to the direction of motion is balanced by the Coriolis force (see Chapter 6).

Thus equation (2.13) can be written:

$$\text{p.g.f.} = -g\,\frac{\partial Z}{\partial n}. \qquad (2.14)$$

The advantage of equation (2.14) is that g, unlike ρ, varies only a little with height. At upper levels equation (2.14) is used extensively in place of equation (2.13).

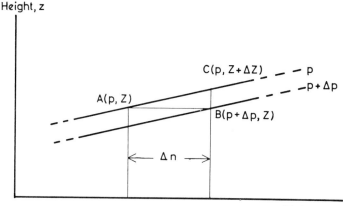

Fig. 2.3. Relation of horizontal pressure gradient $\Delta p/\Delta n$ to the slope $\Delta Z/\Delta n$ of an isobaric surface.

2.3. *Water vapour*

Water vapour is omitted from § 2.1, on the composition of the atmosphere, not because it merits a complete section to itself—which of course it does—but merely to simplify treatment of a system in which it is by far the most significantly varying constituent. This variability arises from the ease with which water substance changes phase under normal atmospheric conditions: no other constituent of the atmosphere has this ability. Changes of phase in themselves, however, would merit little attention were it not for the accompanying changes in latent heat, which for water vapour are particularly large. It follows that the water vapour content of the atmosphere represents a vast storehouse of heat, in latent form. There is a net input to the storehouse in low† latitudes, where evaporation exceeds precipitation and, necessarily, a net output in middle and high latitudes, in the form of excess precipitation. On every scale of atmospheric motion from the hemispherical downwards, the part played by the passage of water substance into and out of its vapour phase is of paramount importance. The maxim that ' in meteorology air can be regarded very largely as diluted water vapour ' should perhaps be a text for all aspiring meteorologists.

† Excluding the equatorial zone itself.

15

2.3.1. *Humidity mixing ratio*

The amount of water vapour in a sample of moist air can be expressed quite simply by its contribution ρ_v to the total density ρ of the sample; e.g. $\rho_v = 15$ g m^{-3}. However, this is not always a helpful parameter in practice, being a function of both pressure and temperature. An alternative indicator is the partial pressure exerted by the water vapour in the sample, referred to simply as its vapour pressure e; but again e varies in proportion to the total pressure p of the sample. In consequence, the humidity mixing ratio[†] r is defined as the ratio of the mass of water vapour in a given sample to the mass of dry air with which it has been mixed to form the sample. Thus if unit volume of a moist air sample contains ρ_v kg of water vapour and ρ_d kg of dry air,

$$r = \frac{\rho_v}{\rho_d}. \tag{2.15}$$

(It is usual to express ρ_v in grammes, so that r has units of g kg^{-1}). So defined, r is a function of water vapour content alone: consequently r is the most straightforward and most generally useful of the various humidity parameters. However, as the partial pressure e and total pressure p are more readily measured than the densities ρ_v and ρ_d it is convenient to express r in terms of e and p.

Consider the gas equation written out for a moist air sample, and again separately for its two components, water vapour and dry air, as follows:

$$p \doteq \rho \frac{R^*}{M_m} . T, \tag{2.16}$$

$$e = \rho_v \frac{R^*}{M_v} . T \tag{2.17}$$

and

$$p - e = \rho_d \frac{R^*}{M_d} . T, \tag{2.18}$$

where M_m and M_d are the mean molecular weights of the moist sample and of dry air respectively and M_v is the molecular weight of water vapour[‡]. From equations (2.15), (2.17) and (2.18) we see that

$$r = \frac{M_v}{M_d} . \frac{e}{p-e}. \tag{2.19}$$

For most practical purposes $M_v/M_d = 18/29 \doteq \frac{5}{8}$ and $e \ll p$, so that

$$r \doteq \frac{5}{8} . \frac{e}{p}; \tag{2.20}$$

e.g. if $e = 16$ mb, $p = 1000$ mb, then $r = 0.01$ kg kg^{-1} = 10 g kg^{-1}.

† Referred to simply as mixing ratio.
‡ M_d is equivalent to $\overline{M}$ in § 2.1.

16

2.3.2. Density of moist air

The mean molecular weight of a moist air sample can be derived from equations (2.16), (2.17) and (2.18), remembering that

$$\rho = \rho_v + \rho_d, \qquad (2.21)$$

or directly from equation (2.4), to give

$$M_m \doteq M_d(1 - \tfrac{3}{5} \cdot r). \qquad (2.22)$$

Thus moist air is lighter than dry air by a small amount proportional to the quantity of water vapour present. Combining equations (2.16) and (2.22), the corresponding expression for the density of a moist air sample is obtained, viz.

$$\rho = \frac{p}{RT} \cdot (1 - \tfrac{3}{5} \cdot r), \qquad (2.23)$$

which, in the limit of zero vapour content, reduces to equation (2.5).

It is sometimes convenient to write equation (2.23) in a form entirely similar to equation (2.5), i.e.

$$\rho = p/RT_v, \qquad (2.24)$$

where T_v, equal to $T/(1 - \tfrac{3}{5}r)$, is called the ' virtual temperature ' of the moist air. This is the temperature at which a fictitious neighbouring volume of dry air would have the same density, and therefore the same buoyancy, as the moist air. If virtual temperature is substituted for (real) temperature in a relationship derived for dry air, such as equation (2.12), then that relationship becomes at once applicable to moist air: this is the value of the virtual temperature concept.

2.3.3. Saturation vapour pressure

At any particular temperature there is in general an upper limit to the partial pressure which any vapour can exert, known as its saturation vapour pressure, or s.v.p., e_s. For water vapour, e_s is defined as the vapour pressure immediately above a plane† surface of pure water, supercooled water, or ice. The symbols e_i and e_w are used for saturation with respect to ice and water (or supercooled water) respectively. (Much atmospheric water is in the supercooled state.) Figure 2.4 shows the s.v.p. curve $e = e_w(T)$, accompanied below 0°C by the corresponding curve $e = e_i(T)$. All air samples existing to the right of the e_w curve are unsaturated with respect to water, while any in the segment XYZ would simultaneously be supersaturated with respect to ice. (In such circumstances ice crystals would grow at the expense of any supercooled water droplets present.) The region to the left of the e_w curve represents supersaturation with respect to a level surface

† The influence of curvature on the value of e_s over a droplet and the effect on e_s of dissolved impurities are discussed in § 4.1.2.

of pure water: such a situation is most unlikely in the atmosphere owing to the abundance of natural condensation nuclei. Thus a typical air sample may be represented on fig. 2.4 by the point $O(T, e)$.

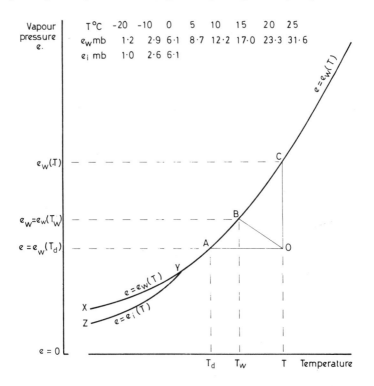

Vapour	T °C	-20	-10	0	5	10	15	20	25
pressure e.	e_w mb	1·2	2·9	6·1	8·7	12·2	17·0	23·3	31·6
	e_i mb	1·0	2·6	6·1					

Fig. 2.4. Saturation vapour pressure versus temperature.

2.3.4. *Paths leading to saturation*

Only if an air sample is saturated can its vapour pressure be determined from a knowledge of temperature alone: for unsaturated air, in contrast, temperature and vapour pressure are independent, in the sense that knowledge of one alone is no guide to the value of the other. If the use of sophisticated laboratory techniques to make an independent determination of e is ruled out, e can be found most easily by bringing the relevant air sample, or part of it, to saturation, in a manner open to simple in erpretation. For example, removal of sensible heat from the sample at 0 in fig. 2.4, under conditions of constant pressure, would result n the sample reaching saturation at the point $A(T_d, e)$ on the s.v.p. curve, where T_d is known as the dew-point temperature. Clearly

$$e = e_w(T_d), \qquad (2.25)$$

18

i.e. the vapour pressure in an air sample is equal to the s.v.p. at its dew point.

The degree of saturation of an air sample is best expressed by its relative humidity U, which may be simply defined as the percentage ratio of the actual vapour pressure in the sample to the maximum permissible vapour pressure at the same temperature, so that

$$U = \frac{e}{e_w(T)} \times 100\%; \qquad (2.26)$$

or, from equation (2.25),

$$U = \frac{e_w(T_d)}{e_w(T)} \times 100\%. \qquad (2.27)$$

Referring again to fig. 2.4, we see that the state $e_w(T)$ would be attained by isothermal addition of water vapour to the sample, at constant total pressure, until the point C on the s.v.p. curve is reached.

There is one other simple manner in which an air sample can be brought to saturation, and that is by the evaporation of water into the sample under adiabatic conditions. Under such conditions the required amount of latent heat is provided by the sample itself, whose sensible heat content (and temperature) must decrease accordingly. Thus the sample would reach saturation at a point $B(T_W, e_w(T_W))$ on the s.v.p. curve, somewhere between the points $A(T_d, e)$ and $C(T, e_w(T))$. The temperature T_W is the adiabatic saturation temperature, or 'thermo-dynamic wet-bulb temperature', of the sample.

A simple relationship between e and T_W exists, and can be derived as follows. Let L_v be the latent heat of vaporization of water (at, or close to, the temperature T_W) while c_p and c_{pv} are the specific heats of dry air and of water vapour, respectively, at constant pressure. Then, if the mixing ratio of an air sample is increased during the process of adiabatic saturation from r to r_W, where $r_W = r_w(T_W)$ the saturation value of r at temperature T_W, it follows from the definition of mixing ratio that

$$L_v(r_W - r) = c_p(T - T_W) + r c_{pv}(T - T_W). \qquad (2.28)$$

Moreover, neglecting the change in heat content of the water vapour initially in the sample (i.e. neglecting $r c_{pv}$ in comparison with c_p) and rearranging terms, we see that

$$r = r_W - \frac{c_p}{L_v}(T - T_W), \qquad (2.29)$$

from which, using equation (2.20):

$$e = e_W - Ap(T - T_W), \qquad (2.30)$$

where $A = c_p/\tfrac{5}{8}L_v$ and $e_W = e_w(T_W)$.

This last equation may be used to find e from a knowledge of T_W and T. However, in meteorology, determinations of vapour pressure by direct measurement of adiabatic saturation temperature are just not made—laboratory conditions and/or techniques being again required. The same applies to direct measurement of dew-point temperature, at least in so far as routine ground-based observations are concerned, the requirement being progressive cooling of a polished surface until deposition of dew is observed.

2.3.5. *Measurement of vapour pressure*

In practice, e is found by a simple employment of two similar thermometers, one ' dry ' and the other ' wet '. The former provides the air temperature T, while the latter, whose bulb is covered by muslin wetted with pure water, provides a wet-bulb temperature T_W' lower than T by an amount depending on the degree of saturation of the air—itself a function of e and T—and on the rate of airflow over the bulbs. Such an arrangement is termed a wet- and dry-bulb hygrometer, or psychrometer if forced ventilation is employed. T_W' is related to but not equivalent to T_W, although for many types of *ventilated* instrument it happens to be equal to T_W†.

There are two distinct reasons why T_W' differs from T_W. Firstly, not all the heat maintaining evaporation from the wet bulb is extracted from air in contact with it—some is conducted along the thermometer stem and some derives from radiation exchange between the wet bulb and its immediate surroundings (thermometer screen; solar radiation shield . . .) which are at a temperature closer to T than to T_W'. Secondly, water vapour and heat diffuse at different rates between the wet bulb and the air immediately around it. Nevertheless, the relationship between e and T_W' may still be written in a form similar to that expressed in equation (2.30) between e and T_W, namely

$$e = e_W' - A'p(T - T_W'),\qquad(2.31)$$

where $e_W' = e_w\ (T_W')$ and A' is constant for (i.e. characteristic of) a given instrument or type of instrument under stated conditions of ventilation—see problem 2.4.

2.3.6. *Distribution of water vapour*

The partial pressure exerted by water vapour at any point in the atmosphere is limited to the saturation value appropriate to the temperature at that point, provided that it does not exceed the local value of total pressure p, in which case e is obviously limited to p. This proviso is of no real concern in the troposphere and lower stratosphere:

---

† For this reason the thermodynamic wet-bulb temperature, T_W, of an air sample is, in meteorology, commonly referred to as its wet-bulb temperature.

however, above about 35 km $e_w(T)$ everywhere exceeds p so that (water) cloud formation is impossible†.

Assuming an average lapse rate of 6·5 K km⁻¹ throughout the troposphere it is easily shown that the value of e_w near the tropopause (at about 10 km) is less than 1% of its value near mean sea level: in contrast p is close to 25% of its sea-level value (fig. 2.1). It follows that the atmosphere's ability to hold water vapour, expressed by the local value of saturation mixing ratio r_w, must decrease rapidly upwards through the troposphere.

It can be shown further (see problem 2.5) that if the troposphere were saturated, or had the same relative humidity at all levels, 75% of all water vapour would lie below about 2·5 km—a figure representative enough of reality even although the vertical distribution of relative humidity is seldom if ever uniform at any given time or place. In comparison, 75% of the total mass of the atmosphere lies below 10 km; thus the strong temperature dependence of e_w results in vapour pressure falling off four times more quickly on average than total pressure, within the troposphere. At higher levels vapour pressure is unlikely to exceed the small values appropriate to the tropopause.

Questions on Chapter 2

2.1. Find the mean molecular weight of dry air composed of the mixture shown in the table.

Gas	N	O	A
% by weight	75·5	23·2	1·3
M kg kmol⁻¹	28·0	32·0	39·9

What would $\bar{M}$ be if 90% of the oxygen and 10% of the nitrogen were dissociated?

2.2. Show that by crediting a past temperature of 1500 K to the Martian exosphere a possible explanation of the present lack of oxygen and abundance of nitrogen on that planet is provided.

$$c = (3R^*T/M)^{1/2} \text{ and } v_e = (2G\mu/r)^{1/2}.$$

G is the constant of universal gravitation, equal to $6·67 \times 10^{-11}$ S.I. units, while μ is the mass of Mars, $6·5 \times 10^{23}$ kg, and r the radial distance to its exosphere, about 3500 km.

2.3. Water vapour makes up 2% by weight of a moist air sample. What is its mixing ratio in g kg⁻¹? What is the vapour pressure if the total pressure is 976 mb? If the relative humidity of the sample is 79% what maximum vapour pressure can it sustain at the same temperature?

† Except, at times, near the temperature minimum around 85 km.

21

2.4. The psychrometer constant A' can be written in the form $A' = A(\kappa/D)^{2/3} (1+f/100)$, where f is the percentage ratio of the quantity of heat reaching the wet bulb by radiation and ' illicit ' conduction to the amount conducted directly from the air. κ and D are the molecular diffusivities in air of heat and water vapour, respectively, while $(\kappa/D)^{2/3} = 0\cdot90$.

Find the value of f appropriate to a wet-bulb thermometer exposed in a thermometer screen, for which $A'p$ is usually assumed to have the value $0\cdot80$ mb K^{-1} at $p = 1000$ mb.

For many types of strongly ventilated psychrometer A' is coincident in value to A. What value of f is then implied?

2.5. Assuming a uniform lapse rate of $6\cdot5$ K km^{-1} throughout the troposphere and likewise a uniform relative humidity, show that. the fractional rate of decrease in vapour pressure with height in the troposphere must be on average about four times greater than the corresponding fractional rate of decrease in total pressure. (Variation in e_w with temperature is given by

$$\frac{1}{e_w} \cdot \frac{de_w}{dT} = \frac{M_v L_v}{R^* T^2}$$

which is equal to

$$\frac{1}{e} \cdot \frac{de}{dT}$$

provided U is constant.)

Show further that under the above conditions (and assuming the average rate quoted) 75% of the water vapour in the atmosphere would lie below $2\cdot5$ km.

CHAPTER 3
heat transfer

ALTHOUGH the atmosphere depends for its heat on radiant energy from the sun it absorbs directly only about one-sixth of the available solar energy. More than one-third of the energy is lost by reflection and back-scattering to space. The remainder—rather less than one-half—is absorbed by the Earth's surface and is then available for heating the atmosphere at second hand.

The atmosphere absorbs a much larger fraction of the Earth's heat radiation than it does of solar radiation. Convection from the surface carries up water vapour as well as heat: when this condenses higher up the latent heat involved is released to the air. The ease with which convection occurs in the troposphere depends on the prevailing lapse rate of temperature.

Land and water absorb and release heat in very different ways. This has important consequences on the weather on all geographical scales.

3.1. *Radiation processes*

3.1.1. *Solar radiation: its energy distribution*

The sun emits continuously and in all directions a stream of electro-magnetic radiation which, unlike the other two methods of heat transfer (conduction and convection), is able to traverse space. The Earth, a body of radius (E) $6\cdot37 \times 10^3$ km, moving in a slightly elliptical orbit round the sun at a mean distance (d) of $1\cdot49 \times 10^8$ km from it, intercepts a minute fraction ($\pi E^2/4\pi d^2 = 5 \times 10^{-10}$) of the total solar radiation output. Ninety-eight per cent of the solar radiation which reaches the fringe of the atmosphere covers a range of wavelengths from the 'ultra-violet' ($0\cdot25$ to $0\cdot4$ μm) through the 'visible' ($0\cdot4$ to $0\cdot7$ μm) to the 'infra-red' ($0\cdot7$ to $3\cdot0$ μm); the remainder of the radiation is at still shorter wavelengths (far ultra-violet, gamma rays. and X-rays) and longer wavelengths (far infra-red and radio waves).

The solar radiation intensities at different wavelengths, outside the atmosphere, had previously to be inferred from measurements made at high-level observatories but they are now directly measurable by satellites. The distribution is rather complex because the energy originates at various solar levels and is subject to some absorption in the solar atmosphere. For our purposes the relative energies of solar radiation per wavelength interval is sufficiently well represented by the theoretical (Planck) distribution for a 'black body' (i.e. a perfect radiator) at 6000 K shown in fig. 3.1; this is referred to as the black-body (radiation) temperature of the sun. The wavelength of maximum intensity of radiation is about 0·5 μm, in the blue-green of the visible; also, since area under the curve between selected wavelengths is proportional to the energy contained in the wavelength interval (both scales being linear), about 45% of the energy is in the visible, 45% is beyond the red of the visible, and the remaining 10% is in the ultra-violet.

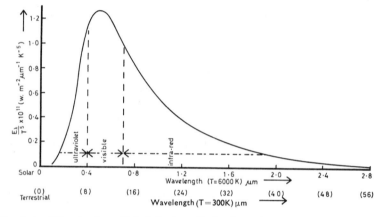

Fig. 3.1. Distribution of black body radiation intensity with wavelength (λ) for solar radiation (T = 6000 K) and terrestrial radiation (T = 300 K): an abscissa scale of λT is the same for all values of T—see § 3.1.4.

The energy curve at the Earth's surface is slightly different from that outside the atmosphere because the atmospheric constituents scatter or absorb some wavelengths more intensely than others; in particular, no ultra-violet radiation of wavelength less than 0·29 μm reaches sea level.

3.1.2. *The solar constant*

If the sun did not vary in its radiation the intensity at the boundary of the atmosphere would depend only on the small annual changes of sun–Earth distance which are associated with the eccentricity of the Earth's orbit; sun and Earth are nearest at perihelion on 1 January and farthest apart at aphelion on 1 July, the maximum difference from the average distance being ± 1·7%. The fundamental quantity relating to solar radiation is then the amount of energy, of all wavelengths, incident in unit time on unit area of a surface placed at right angles to the sun's rays at the Earth's mean distance from the sun. This quantity is called the 'solar constant'. The

24

main difficulty in its measurement from the ground is in making appropriate allowance for the extinction of the radiation in its passage through the atmosphere; as previously mentioned, this is different for different wavelengths. The numerical value of the solar constant is 1.39 kW m^{-2}.

Whether or not the solar constant is, in fact, constant has long been a matter of controversy. Although the sun is 'steadier' than most visible stars it nevertheless varies in its activity. Sunspots provide the most obvious sign of its variability; these are visible to the naked eye and their waxing and waning with a period of about 11 years has been documented for some centuries. Visible evidence includes also solar flares and prominences and other features familiar to astronomers. The variations in the intensity of solar ionizing radiation are known to affect the lowest part of the ionosphere at about 70 km. It is less certain that solar variability affects the ozone layer at 20–50 km and very doubtful whether still lower atmospheric levels are affected.

Possible connections between solar activity on the one hand, and the solar constant and weather phenomena on the other, have long been debated. The present consensus view is that there must, in principle, be variations in the solar constant which are too small to be identified by present measurements. The variable component of solar radiation is almost entirely in the far ultra-violet. Its effects are confined to the high atmosphere where it is completely absorbed. The weather elements at low levels are not influenced by solar activity.

The Earth, rotating under the stream of solar radiation, has a surface area of $4\pi E^2$ ($E =$ Earth's radius), while the cross-sectional area which intercepts the stream is πE^2. Thus the average intensity of radiation perpendicular to the Earth's surface is one-quarter of the solar constant. It is shown in the next section that not all this available energy is effective in heating the atmosphere.

3.1.3. *Effect of the atmosphere and Earth on solar radiation*

When radiation traverses a medium the amount absorbed depends on the radiation intensity, on the amount of substance traversed and on the ability of the substance to absorb the radiation. In the case of solar radiation traversing the atmosphere, the radiation covers a wide range of wavelengths and the medium consists of a large number of gases, each capable of absorbing (if at all) in bands of wavelengths individual to itself (though sometimes overlapping those of other constituents). The atmospheric constituents also scatter a proportion of the radiation in all directions while some of them, and also the Earth's surface itself, reflect part of it.

(1) *Scattering*

When a beam of radiation, such as that from the sun, encounters very small particles none of the original energy is lost but it is diffused in all directions from the particles. The chief scattering agents in the atmosphere are air molecules, water vapour and dust. They return to space an estimated 6% of the incident radiation. A further 20% reaches the Earth's surface from all directions after multiple scattering; this is the ' diffuse ' or ' sky ' radiation (as opposed to the ' direct ' solar radiation) which illuminates all parts of the sunlit hemisphere which are not in direct sunlight.

Scattering of sunlight produces various optical effects. The intensity with which air molecules scatter radiation varies inversely with the fourth power of the wavelength λ; this is termed ' Rayleigh scattering ' and applies to particles of radius $< 10^{-1}\,\lambda$. The effect of Rayleigh scattering is best seen when the atmosphere is (almost) free of suspended particles. The sky light is then a brilliant blue and distant hills may acquire a blue tinge owing to the high intensity of light scattered by air molecules in the long sight-path; on the other hand, when a low sun is viewed directly it appears red because mainly blue light has been scattered out of the direct beam. The effect of dust and haze in the atmosphere is to whiten both the sky light and the landscape because the intensity of scattering from such ' large ' particles (radius $> 25\,\lambda$) is independent of wavelength. Exceptionally, but notably, a local source of suspended particles of an intermediate and sufficiently uniform size (of the order of $1\,\mu$m) causes the sun (or moon) to appear blue because this size of particle scatters mainly red light from the solar beam; hence the saying ' once in a blue moon '. Examples of such local sources have included exceptional duststorms, forest fires and volcanic eruptions.

(2) *Reflection*

The brilliant white of a fresh snow surface in sunlight indicates that most of the light is being reflected and that very little of it is absorbed. In contrast, a tropical ocean under a high sun reflects hardly any of the direct sunlight and only a small part of the sky light; nearly all the light is absorbed. This contrast in behaviour underlines the important question—what happens to the solar radiation which reaches different types of surface?

The ratio of the intensity of solar radiation reflected by a surface to that incident on it is called the albedo of the surface. The proportion of radiation reflected varies somewhat with wavelength but the term is usually applied to solar radiation as a whole and is ideally measured by radiometers which respond to all solar radiation wavelengths. The total albedo is closely related to visual albedo, as may be seen from the following approximate values: fresh snow 90%, old snow 50%, vegetation 20%, light-coloured soil 30%, dark-coloured soil 10%, clouds 50–90% depending on type, water surfaces 10% (but much higher when the sun is very low). These values are only approximate since there are variations with such factors as surface roughness, soil type and soil moisture content.

The average albedo for the whole Earth and atmosphere depends mainly on the variations of cloud amount and snow cover with season and latitude. It is estimated that 24% of total radiation is reflected from clouds and a further 6% from the Earth's surface. When the 6% of radiation which is scattered from the atmosphere back to space is added, a planetary albedo of 36% is obtained. Satellite measurements indicate a rather lower value than this, while a slightly higher value (of *visual* albedo) has been obtained from comparisons of the light (sun-lit) and/dark (Earth-lit) segments of the moon.

(3) *Absorption*

Absorption of radiation by a substance causes its temperature to rise to a level at which it loses heat (by one or other of the heat transfer processes) at a rate which just balances the rate at which heat is gained by absorption. If, therefore, we refer back to fig. 2.1 we see that absorption of solar radiation occurs in three main regions, namely in the thermosphere, near the stratopause at 50 km and at the Earth's surface†.

In the thermosphere, warming is by absorption of wavelengths below 0·18 μm (mainly by molecular oxygen) these wavelengths being completely absorbed above 85 km. Below this level, radiation of wavelengths near 0·2 μm dissociates oxygen molecules into two oxygen atoms, and the oxygen molecules and atoms combine to form ozone (O_3) molecules. The rate of ozone production at first increases with depth of penetration as the dissociating radiation meets denser gas; but it then decreases and finally ceases at still lower levels because the wavelengths concerned have been absorbed higher up. A third process, involving ozone destruction by its absorption of wavelengths of 0·2 to 0·3 μm, competes with the other two processes. The absorption by ozone is so intense that little of this radiation penetrates much below 40 km. The sharp temperature maximum at about 50 km (see fig. 2.1) is attributed to absorption by this gas. (Ozone is moved downwards from its main level of formation to a lower stratospheric level where it is not destroyed by the ultra-violet radiation.)

Apart from the loss of about 3% in the high atmosphere, a further 14% of total radiation is absorbed, mainly by water vapour and clouds at low levels. Cloud absorption is much the less important since clouds are poor absorbers, as shown by the ability of a sheet of thin low cloud or fog over the sea to withstand strong sunshine. Since 36% of the radiation is lost to space the remainder (47%) is absorbed at the

† Owing to the great differences of density at these levels, the mean temperatures do not of course reflect the respective *amounts* of solar radiation absorbed.

Earth's surface which is therefore the atmosphere's main effective heat source. The solar energy budget for the Earth and atmosphere is summarized in fig. 3.2.

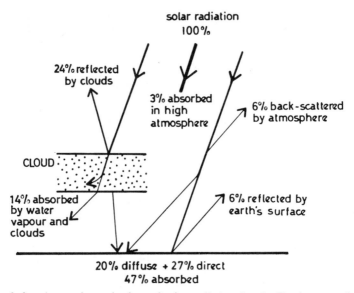

Fig. 3.2. Approximate budget of solar radiation for the Earth–atmosphere system.

3.1.4. *Radiation from the Earth and atmosphere*

(1) *General laws*

Every body (at a temperature above absolute zero) radiates heat at a rate which depends on its nature, surface area and temperature.   Its efficiency as a radiator is judged by comparing its rate of emission per unit area with that of a ' black-body ' under the same conditions.   All terrestrial objects, including clouds, have an efficiency of between 90% and 100%; for practical purposes they are treated as black-body radiators, each at its own temperature.

The temperature of the radiator is vital in several respects.   First, the lower the temperature of a body the longer are the wavelengths of its radiation; a familiar illustration of this is the progressive cooling of strongly heated metal from ' white heat ' to ' red heat ' to invisible infra-red.   A change of radiator temperature causes no change of *shape* of the energy–wavelength curve.   However, the wavelength limits do change; so also does the wavelength of maximum intensity (λ_{max}) in accordance with Wien's displacement law, $\lambda_{max}T = 2.88 \times 10^{-3}$ m K, where T K is the temperature of the radiator.   The changes in the axes scales required to accommodate different radiator temperatures are

28

shown by referring back to fig. 3.1; the wavelength scale varies inversely with T and the energy scale varies directly as T^5. The scale for $T = 300$ K (a typical low-latitude surface temperature) is shown in fig. 3.1 to peak at about 10 μm. Since it overlaps hardly at all in wavelength with the solar radiation curve the two streams are often termed 'long-wave' and 'short-wave' radiation, respectively.

A further important effect of differences of temperature of radiating bodies is expressed in Stefan's law $E = \sigma T^4$, where E is the intensity of radiation per unit area of a black-body at temperature T K, and σ (Stefan's constant) $= 5 \cdot 70 \times 10^{-8}$ W m^{-2} K^{-4}.

Finally, a further radiation law (Kirchoff's) states that if a body absorbs certain wavelengths well it also emits these well; and that wavelengths which are not absorbed are not emitted. For example, a snow surface radiates long waves with almost as much intensity as a perfect radiator and absorbs nearly all the long-wave radiation emitted by trees, buildings, clouds, etc; on the other hand, it absorbs very little short-wave radiation and emits hardly at all in these wavelengths.

(2) *Atmospheric absorption and exchange*

The intensities of the absorption bands of the atmospheric gases, the wavelengths of these bands relative to the long-wave spectrum, and the amounts of absorbing gases jointly determine how much of the long-wave radiation is absorbed in the atmosphere. It turns out that nearly all the absorption is by certain minor constituents, namely water vapour, carbon dioxide and ozone. Water vapour dominates the absorption in the troposphere, carbon dioxide in the mesosphere; all three are significant in the stratosphere. The radiation is emitted upwards and downwards between the absorbing gases and is passed eventually, some of it directly, down to earth (to be re-radiated there) or out to space. Cloud intercepts all infra-red radiation and re-radiates it up and down with an intensity which depends on the temperatures of its top and base.

The infra-red radiation exchanges in the troposphere do no more than modify slightly the temperature lapse rate which, as will be explained later, is mainly determined in this region by convection from the surface†. An essential difference is that, whereas the temperature of all the constituent gases of the air are changed by vertical motion, only some minor constituents—usually less than 1% of the total—participate in the radiation exchanges. Above the tropopause vertical motion is much less common; but where it occurs it modifies the lapse rates, which are determined in these regions mainly by radiation exchanges. In the following account, positive lapse rates of both temperature and humidity in the troposphere will be assumed.

† Polar latitudes in winter, and low atmospheric levels elsewhere on 'radiation nights', are exceptions.

The approximate absorption curves for water vapour, carbon dioxide and ozone are shown in fig. 3.3 in relation to the radiation curve for 288 K. The region from about 8 to 12 μm is the 'atmospheric window' through which some of the surface radiation passes directly to space unless intercepted by cloud. (Ozone has an absorption band at about 9·6 μm which absorbs part of the radiation in this region.) The fraction of the surface radiation which passes directly out, averaged for the Earth as a whole, depends on the percentage cloudiness (about 54% overall), including its latitudinal variation; the fraction is estimated to be about 10%.

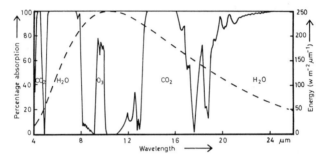

Fig. 3.3. Percentage absorption by water vapour, carbon dioxide and ozone of terrestrial radiation for $T = 288$ K (after Gates).

The passage to space of radiation of absorbed wavelengths may be considered to take place from layer to layer, each layer being deep enough to absorb these wavelengths completely and re-radiate them in all directions with an intensity which depends on the layer temperature. The depth of layer required is controlled by the amount of moisture available since there is always enough carbon dioxide for full absorption in its wave bands. The lowest fully-absorbing layer is shallow but the layers become progressively deeper upwards as humidity decreases. Owing to the decrease of temperature with height, each layer sends downwards less radiation than it receives from below (Stefan's law). There is thus a net stream of radiation upwards to space. The highest fully-absorbing and emitting layer is within the troposphere, but there is still significant emission to space from the stratosphere and mesosphere.

Equilibrium may be assumed between the streams of solar and infra-red radiation absorbed and emitted by the Earth and atmosphere since (almost) no long-period change of their temperature has been observed. Thus the mean temperature (T) at which the Earth (radius E) and its atmosphere radiate to space is given by

$$4\pi E^2 \sigma T^4 = \pi E^2 (1 - A)S, \qquad (3.1)$$

where σ is Stefan's constant and S the solar constant; for $A = 0.36$

30

this gives $T = 250$ K. T is the temperature at which the high atmosphere radiates to space, weighted by minor contributions from the temperatures of the Earth's surface and cloud tops because of their emission through the atmospheric window. $T = 250$ K would be the Earth's mean surface temperature if the atmosphere and its clouds did not absorb infra-red radiation; this compares with the actual value of about 288 K—a difference of nearly 40 K.

The atmosphere's (relative) transparency to solar radiation and its 'blanketing' of the Earth's radiation are also seen in the following comparison. Short-wave radiation absorbed by the Earth's surface averages $0.47\,S/4 = 0.166$ kW m^{-2}, while long-wave radiation from the surface is $\sigma T^4 = 0.392$ kW m^{-2} (for $T = 288$ K), i.e. the surface emits more than twice as much energy in the infra-red as it receives in solar radiation. The atmosphere—especially clouds and water vapour at lower levels—returns to the Earth much of the energy radiated by the Earth and, in addition, emits downwards part of the 17% of the solar energy which it absorbs directly.

(3) *Radiation balance*

Measurements and calculations show that, when allowance is made for the return radiation from the atmosphere, the net long-wave radiation from the surface amounts, on average, to less than half the solar radiation which is absorbed there, i.e. the Earth's surface has a large positive radiation balance. The balance is greatest in the tropics and changes sign near the poles—see fig. 3.4. The result is simply explained by the fact that heat and, even more important, water vapour are convected from the surface into the troposphere. This heat, and also the latent heat released when the water vapour condenses, are

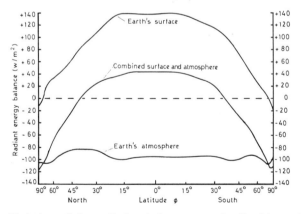

Fig. 3.4. Variation of the radiation balance over the Earth's surface (after Sellers). Since energy is plotted against $\sin\phi$, unit area on the diagram corresponds to energy per unit area of the Earth's surface.

31

subsequently radiated from the troposphere, in addition to the 14% solar radiation absorbed there; the troposphere thus radiates more energy than it absorbs (radiation balance negative).

Since the temperature inversion at the tropopause acts as a barrier to the convection of heat to higher levels (see § 3.2.5) the amounts of radiated heat absorbed and emitted by gases in the mesosphere and stratosphere should be about equal ('radiative equilibrium'). This is verified by calculation, provided averages in space and time are considered.

The radiation balances for the atmosphere alone, and for the Earth-atmosphere system (positive below 40°, negative elsewhere) is again explained by motion. Winds move warm air and water vapour from low to high latitudes and cold air in the opposite direction. Ocean currents also play a part in the heat transfer.

(4) *Atmospheric heating and cooling*

Where ΔH is the energy absorbed in a unit air column of mass m and specific heat c_p, the corresponding temperature rise ΔT is given by $\Delta H = mc_p \, \Delta T$; $m = 10 \cdot 2 \, \Delta p$ kg m^{-2}, where Δp(mb) is the pressure difference between the top and bottom of the column, $c_p = 10^3$ J kg^{-1} K^{-1} and ΔH is the energy (joules) supplied to 1 m^2 cross-section. The average daily rate of heating caused, throughout an entire atmospheric column of 1000 mb, by absorption of 17% of the available solar radiation $(0 \cdot 17 \times S/4)$ is found from this formula to be $0 \cdot 5$ K day^{-1}. Similarly, the average cooling by emission of long-wave radiation is about $1 \cdot 3$ K day^{-1}. Thus the atmosphere (almost exclusively the troposphere) cools at an average daily rate of about $0 \cdot 8$ K by radiative exchange.

3.2. *Convection*

The upward transfer of heat through the atmosphere by conduction, in the strict sense (i.e. by the action of molecules of greater energy on those of less), is important only in the very shallow layer of air—perhaps 1 mm in depth—which adheres to the Earth's surface. Above this layer the effect of conduction is negligible compared to the effects of radiation and convection.

Convection involves the vertical interchange of masses of air, each carrying its store of heat, water vapour and momentum. The convection may be 'free' (or 'natural') as when the density of a mass is different from its surroundings and it is moved by buoyancy forces; or it may be 'forced', as by the passage of air over rough ground. Commonly it is a combination of both. The motion is in discrete masses or 'eddies' of a wide variety of size; other terms employed are 'particles', 'parcels', and (restricted to free convection from the surface) 'thermals'.

Although convected masses mix to some extent with their

surroundings while they are moving vertically, the simplifying assumption that there is no mixing gives useful results on lapse rates and stability. This assumption will be made in this section.

3.2.1. *Adiabatic temperature changes*

When a liquid is stirred its temperature is made uniform: when a gas is stirred a temperature lapse is established within it. This difference is caused by the very different compressibilities of the two types of fluid. A rising air parcel expands because the inward pressure on it, exerted by the surrounding air, decreases with height; the work done by the parcel in expanding is at the expense of its internal energy and its temperature falls. Conversely, the atmosphere does mechanical work on a descending parcel in compressing it; its internal energy therefore increases and its temperature rises. Vertical motion is usually rapid enough to make these temperature changes far outweigh the effects of heat interchange by conduction and radiation, between the parcels and their surroundings. The changes are therefore (nearly) adiabatic.

3.2.2. *Adiabatic equation*

The above physical reasoning suggests that in an adiabatic temperature change the fundamental relationship is one between T and p. We derive this, starting with required background information.

(*a*) Where a quantity of heat dq supplied to unit mass of a gas produces a temperature change dT the specific heats at constant pressure (c_p) and at constant volume (c_v) are defined by:

$$c_p = \left(\frac{dq}{dT}\right)_p \tag{3.2}$$

and

$$c_v = \left(\frac{dq}{dT}\right)_v \tag{3.3}$$

(*b*) Since pressure (p), specific volume (v), and temperature (T) are related by $pv = RT$, the relationship after small increments dp, dv and dT is

$$(p + dp)(v + dv) = R(T + dT).$$

Multiplying, and ignoring products of increments, we have

$$pdv + vdp = RdT. \tag{3.4}$$

(*c*) The work done by unit mass of a gas at pressure p in expanding by volume increment dv is

$$dw = pdv. \tag{3.5}$$

(work done = force × distance, for unit cross-section).

33

The first law of thermodynamics states that the quantity of heat (dq), supplied to unit mass of a gas, is balanced by an increase in the internal energy of the gas (du) and the external work done by the gas (dw); it is a statement of the principle of the conservation of energy. In symbols,

$$dq = du + dw = du + pdv \quad \text{from equation (3.5).}$$

If there is no volume change (d$v = 0$) then

$$dq = du; \text{ and } c_v = \left(\frac{dq}{dT}\right)_v$$

$$= \frac{du}{dT};$$

$$\therefore \quad du = c_v dT.$$

Thus

$$dq = c_v dT + pdv$$

$$= c_v dT + RdT - vdp \quad \text{from equation (3.4).}$$

If there is no pressure change (d$p = 0$) then

$$dq = (c_v + R)dT \quad \text{and} \quad c_p = \left(\frac{dq}{dT}\right)_p$$

$$= c_v + R,$$

so that

$$c_p - c_v = R. \tag{3.6}$$

Thus

$$dq = c_v dT + pdv$$

$$= (c_p - R)dT + pdv \quad \text{from equation (3.6)}$$

$$= c_p dT - RdT + pdv; \text{ and from equation (3.4)}$$

$$dq = c_p dT - vdp \tag{3.7}$$

If the change is adiabatic, then d$q = 0$ and

$$c_p dT = vdp$$

$$= \frac{RTdp}{p};$$

$$\therefore \quad \frac{dT}{T} = \frac{Rdp}{c_p p};$$

$$\therefore \quad \log T = \frac{R}{c_p} \log p + \text{constant};$$

$$\therefore \quad T = \text{constant} \times p^{R/c_p}$$

$$= \text{constant} \times p^{\kappa}.$$

Now
$$\kappa = R/c_p$$
$$= (c_p - c_v)/c_p$$
$$= (\gamma - 1)/\gamma,$$
where
$$\gamma = c_p/c_v = 1\cdot 4 \text{ for air.}$$
Thus
$$\frac{T}{p^{0\cdot 288}} = \text{constant.} \tag{3.8}$$

3.2.3. *Potential temperature: dry adiabatic lapse rate*

The temperature of an air parcel which is subject to adiabatic expansion or contraction is seen from equation 3.8 to be a function of pressure alone. This equation makes it possible to 'label' an air particle, which is at an arbitrary pressure, by the temperature it would attain if brought adiabatically to 1000 mb. This, the 'potential temperature' (θ), is an effective label whether or not the particle is moved adiabatically.

$$\frac{T}{p^{0\cdot 288}} = \frac{\theta}{1000^{0\cdot 288}};$$
$$\therefore \quad \theta = T\left(\frac{1000}{p}\right)^{0\cdot 288}. \tag{3.9}$$

In an atmosphere in which θ is constant with height the temperature at any level is exactly that which leads, by adiabatic ascent or descent, to the same value of temperature at 1000 mb. The atmosphere is then said to be in a state of adiabatic (or convective) equilibrium. The corresponding lapse rate is established by ascending and descending masses and its value is easily determined, as follows.

For an individual parcel of unit mass which is thermally insulated from its surroundings $dq = 0$ and equation (3.7) takes the form

$$c_p dT - v dp = 0.$$

On substituting for dp from equation (2.7),

$$c_p dT + v\rho g dz = 0,$$

which simplifies, with $v\rho = 1$, to

$$c_p dT + g dz = 0;$$
$$\therefore \quad \frac{dT}{dz} = -\frac{g}{c_p} \tag{3.10}$$
$$= -9\cdot 8/10^3 \text{ S.I. units,}$$

and the lapse rate
$$\frac{dT}{dz} = 9\cdot 8 \text{ K km}^{-1}.$$

35

For positive dz (rising mass) the temperature change is negative, while for negative dz (falling mass) it is positive. This lapse rate applies to unsaturated parcels and is called the dry adiabatic lapse rate (DALR), designated Γ. The above numerical value holds, strictly, if at all levels the parcel is surrounded by air at its own temperature. (This is implicit in the cancellation of the produced $v\rho = 1$ because v applies to the parcel and ρ to the environment.) For a parcel of temperature T' surrounded by air of temperature T the correction factor T'/T must be applied to the lapse rate in equation (3.10). In practice, this correction is negligible because T and T' are in K and $(T - T')$ does not exceed a few degrees.

3.2.4. Saturated adiabatic lapse rate

Adiabatic cooling by ascent, if sufficiently prolonged, leads to saturation of even the driest of natural air parcels. Further ascent is at a rate less than Γ because the cooling of a parcel by expansion is then offset, to some extent, by the latent heat released by condensing water vapour: (in the atmosphere saturation results in condensation.) If, in the ascent of saturated air, the saturation mixing ratio r_w changes by dr_w (a negative quantity) owing to condensation, the heat released to unit mass of air is d$q = -L\mathrm{d}r_w$, where L is the latent heat of vaporisation.

Thus,
$$-L\mathrm{d}r_w = c_p\mathrm{d}T - v\mathrm{d}p \qquad \text{from equation (3.7),}$$
so that
$$-\frac{\mathrm{d}T}{\mathrm{d}z} = -\frac{v\mathrm{d}p}{c_p\mathrm{d}z} + \frac{L\mathrm{d}r_w}{c_p\mathrm{d}z}$$

$$= \frac{vg\rho}{c_p} + \frac{L\mathrm{d}r_w}{c_p\mathrm{d}z} \qquad \text{from equation (2.7),}$$
i.e.
$$-\frac{\mathrm{d}T}{\mathrm{d}z} = \Gamma + \frac{L\mathrm{d}r_w}{c_p\mathrm{d}z} \quad \text{since } v\rho = 1. \qquad (3.11)$$

Thus the saturated adiabatic lapse rate (SALR), denoted γ_s, is less than the DALR by an amount directly proportional to the mass of water vapour condensed into the rising air over a fixed height interval; this, in turn, depends on the temperature. At high temperatures, where saturated air contains a large amount of water vapour, γ_s is reduced to about $\Gamma/3$, whereas at low temperatures there is little difference between them.

3.2.5. Stability and instability

Such factors as surface cooling or large-scale subsidence of the atmosphere cause big departures—usually temporary and restricted in depth—from the normal tropospheric lapse rate of temperature. We now examine how different lapse rates inhibit or promote convection in the atmosphere.

(1) *Unsaturated rising or falling mass*

Two environment curves are shown in fig. 3.5, with lapse rates respectively greater than Γ (I) and less than Γ (II); they extend over an arbitrary depth of 2 km and are assumed to be unsaturated. Consider first environment I. If a particle of air at A is displaced upwards its temperature changes at the rate Γ; its temperature (T') is then higher than that of its immediate surroundings (T). Since the pressure on both is the same, the density of the particle is less than that of the environment. It has 'positive buoyancy' and accelerates upwards.

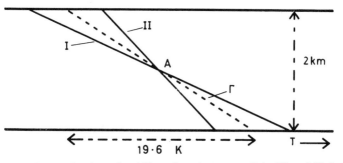

Fig. 3.5. Determination of stability of environments I ($>\Gamma$) and II ($<\Gamma$).

Similarly, a particle which is displaced downwards warms at the rate Γ; it has a lower temperature and higher density than I, its buoyancy is negative and it accelerates downwards. These vertical accelerations are found by equating the net upward force per unit mass (up-thrust minus weight) to the acceleration (a). The result is

$$a = \frac{(T' - T)g}{T}. \tag{3.12}$$

An atmosphere through which unsaturated particles move readily up and down is said to be unstable with respect to them. Its distinctive feature is that its temperature lapse rate exceeds Γ; this is the same as saying that θ decreases upwards in it. The unstable state is normal in the layer above a surface which is kept warm. However, Γ represents (nearly) the upper limit at higher levels because, if it is exceeded, the vertical interchange of masses which is thus encouraged quickly re-establishes the rate Γ.

If the previous argument is applied to environment II it is seen that an air particle (at A, say) which is moved upwards or downwards becomes subject to a force which tends to restore it to A. Thus II inhibits vertical exchange of unsaturated parcels and is said to be stable with respect to them. Finally, an environment whose lapse rate is Γ is in a neutral stability state; θ is the same at all heights and the condition is that of convective equilibrium previously discussed.

The analogy previously drawn between the temperature of a liquid and the potential temperature of the atmosphere (constancy of each established by mixing) thus extends to the condition for instability (decrease of each with height). However, whereas density increases upwards in an unstable liquid (T decreases upwards) the same is not necessarily true of an unstable atmosphere. The critical atmospheric lapse rate for constant density with height is easily found by differentiating the equation $\rho = p/RT$ with respect to z and finding $-dT/dz$ for the condition $d\rho/dz = 0$. Its numerical value (g/R) is 3.4 K km^{-1} (i.e. about $3.5\,\Gamma$); it is of no special significance in the atmosphere, though it is often exceeded close to a strongly heated land surface.

(2) *Saturated rising mass*

Since a saturated mass cools on ascent, not at the rate Γ, but at the lower rate γ_s, it is easily seen that an environment of lapse rate greater than γ_s (but less than Γ) will aid the ascent of saturated masses through it, while resisting the ascent of unsaturated masses. This state—a very common one in the atmosphere—is termed conditional instability. Various complications arise in connection with it. Firstly, such an atmosphere is usually stable for small vertical displacements of parcels but is unstable if the displacements are large enough to cause the parcels to become saturated. Secondly, energy is gained by saturated parcels as they move up through a conditionally unstable atmosphere but is lost by the interchanging parcels because they mostly warm at the rate Γ as they move down, even though initially saturated. Finally, the buoyancy of a rising saturated parcel increases with the dryness of the environment through which it rises; this is because of the (small) dependence of air density on mixing ratio mentioned in § 2.3.2. The analysis of atmospheric stability may thus be quite complex. Examples are discussed in Chapter 5 where a further type of instability, associated with the forced ascent of an atmospheric layer, is described.

3.3. *Heat transfer in land and sea*

Both land and sea act as large heat reservoirs for the atmosphere in that their surface layers absorb heat during the day and during summer and subsequently release it to the surface, and thence to the air, during the night and winter. In this way they mitigate the heat of day and summer and also the cold of night and winter. The sea is much the more powerful in its modifying influence and this has important effects on the geographical and time distributions of many of the weather elements. Their different actions are explained mainly by the difference in the methods of heat transfer within them: in the earth the transfer is by pure conduction while in water it is very largely by convection.

3.3.1. *Heating and cooling of soil*

The depth of soil reached by the daily cycle of solar heating, and also the temperature rise within the top layers, depend on the thermal conductivity of the soil and on its heat capacity (product of density and specific heat). If, for example, the conductivity and heat capacity are both low the heat does not penetrate far and there is a large temperature rise within the surface layer (top few centimetres). In such a case the land's reservoir action is almost absent and temperature falls quickly at night to a low level if atmospheric conditions favour a net radiative loss from the surface.

The air content of a soil is important since air is a poor conductor and has a low heat capacity. Thus the surface layer of a dry soil is warmer during the day, and colder during the night, than that of a wet soil of the same type. A snow surface, and the air in contact with it, reach very low temperatures on a 'radiation night'; in this case its high albedo and the loss of energy in evaporation prevent the surface temperature from rising much during the day. Ground which is covered by vegetation is protected from direct solar rays and therefore has a low daily range of temperature.

In all soils the amplitude of the diurnal wave of temperature decreases rapidly with depth and the times of maximum and minimum are progressively later as the depth increases. With strong surface heating of suitable ground the wave is discernible down to about 0·5 m, although usually to much less than this. The behaviour of the annual temperature wave is similar. Since, however, the depth of penetration of a periodic wave which is transferred by conduction is known by theory to be proportional to the square root of the period, the annual wave may be discernible down to about $0·5 \times \sqrt{365} \doteqdot 10$ m; again it is usually only a small fraction of this. At maximum depth the highest and lowest temperatures are delayed by months compared to those at the surface and the two waves may even be completely out of phase.

3.3.2. *Heating and cooling of water*

Various factors restrict the temperature rise of a water surface exposed to sunshine. First, a substantial fraction of the energy is used in evaporating water from the surface; second, much of the radiation penetrates a long way before it is completely absorbed; third, water has a high specific heat; fourth and most important, movement of the surface water by wind mixes the absorbed heat through a deep layer, and also surface water which is cooled by radiation is replaced by warmer water from below. All these effects combine to make the diurnal temperature variations of a water surface (and of the air in contact with it) very small indeed. Similarly its annual variations, though significant, are much smaller than those over land.

The differences of heat transfer within land and sea establish pressure fields and air motions which then mitigate the effects of these differences (compare § 3.1.4). Thus in summer cool sea air invades warm continents, and in winter cold continental air moves over the warm oceans; these are the monsoon winds. Their counterpart on a diurnal time scale are the land and sea breezes established along coast lines.

Questions on Chapter 3

3.1. What are the percentage variations of solar radiation, relative to the solar constant, received by the Earth at the times of perihelion and aphelion, the percentage variations in the mean sun–Earth distance being $\pm 1.7\%$? How do the annual totals of solar radiation received outside the Earth's atmosphere compare at corresponding latitudes in the two hemispheres?

3.2. Calculate the average intensities of the following types of radiation received at the moon's surface: (a) terrestrial radiation; (b) solar radiation reflected by the Earth; (c) direct solar radiation. What is the moon's radiative equilibrium temperature? Moon–Earth distance $= 3.84 \times 10^5$ km; moon's radius $= 1.74 \times 10^3$ km; moon's albedo $= 0.07$; see text or preliminary pages for other data.

3.3. The estimated average daily evaporation (and precipitation) rate for the Earth's surface as a whole is about 2·7 mm. What percentage of solar radiation, intercepted by the Earth, does this represent?

3.4. What is the potential temperature of air of pressure 600 mb and temperature $-20°C$? At what temperature will it reach the pressure 800 mb? If, in ascending rapidly from 600 mb, the particle reaches a temperature of $-31°C$ at 500 mb what can you say about the nature of the ascent?

3.5. Find the upward acceleration of a saturated parcel which has a temperature excess of 1 K with respect to a saturated environment of temperature 10°C and mixing ratio 6 g kg^{-1}, assuming no mixing. What is the acceleration if the environment relative humidity is 50% and the temperature unchanged?

condensation and precipitation

4.1. *Microphysical processes*

IN this section we examine how condensation of water vapour is produced within air which has been brought to the saturation point; also how precipitation elements develop from the cloud particles and how thunderstorm electricity is generated. This branch of meteorology has probably a closer connection with laboratory physics than any other. Some of the atmospheric processes involved can be simulated there to a certain extent. Also, laboratory techniques have been widely used to determine some of the properties of samples of air, and of clouds and precipitation elements, collected from aircraft specially flown for the purpose.

4.1.1. *Condensation nuclei*

If air, contained in a closed glass chamber, is suddenly expanded by a controlled amount a cloud is seen to form when the relative humidity is (just over) 100%. If a similar air sample is allowed an appreciably greater expansion, so that the air becomes highly supersaturated, the cloud which forms is much denser than before. If the cloud droplets are allowed to settle and the air brought back to its original volume, then again expanded, any cloud that forms is more tenuous. On further repetition the stage is eventually reached when a supersaturated expansion of several hundred per cent produces no condensation of the water vapour present.

The explanation of these events is that air contains a large number of minute particles, each of which provides a surface on which condensation of water vapour may occur. Some of these are effective when the air becomes just saturated, while many others require appreciable supersaturation. In the final stage all the condensation nuclei have been removed from the air.

Atmospheric nuclei originate, for example, in sea-salt spray, chemical smokes, combustion and ordinary dust. They range in size from about 10 μm (rarely) to 10^{-3} μm radius. They number billions per cubic metre, the smaller being much more numerous than the larger.

In the atmosphere, as in the first of the controlled expansions in the cloud chamber, condensation usually occurs on the larger nuclei when the relative humidity is (just over) 100%. Atmospheric cooling by the ascent of air is slow enough to enable these larger nuclei to 'mop up' excess water as it is made available. Supersaturation greater than a fraction of 1% is therefore not observed and the numerous smaller particles (condensation on which occurred in the supersaturated expansions of the chamber air) are not activated in nature. Thus cloud droplets number only a small fraction of all the nuclei present.

4.1.2. *Curvature and solute effects*

Condensation at saturation is not as obvious as might appear. The state of saturation applies to a plane water surface (see § 2.3.3), the s.v.p. for which is less than that over a curved surface such as a droplet. For very small nuclei the s.v.p. is much (up to several times) greater than for a plane surface and so water vapour does not collect on them in the atmosphere. For large nuclei, however, the difference between the s.v.p. values is small and easily bridged, some of the natural nuclei being hygroscopic. Each particle of this kind forms with water a solution over which the relative humidity is less than 100%, so that the 'curvature effect' and the 'solute effect' largely cancel. Occasionally, in industrial regions, a high concentration of large, hygroscopic nuclei builds up near the Earth's surface and a water-droplet fog forms when the relative humidity is well below 100%.

4.1.3. *Water-droplet clouds*

The typical cloud contains about 10^9 water droplets per cubic metre, their radii ranging from about 1 to 20 or 30 μm, the average being about 10 μm. Thus, though the droplet concentration is large, their size is so small that, between any two of them, there is enough room for about 50 others, on average.

Each particle in the atmosphere is subject to the force of gravity and accelerates downwards, relative to the air, until the air's frictional retarding force just balances the gravitational force; the particle then moves at a constant speed called its terminal velocity. The bigger the particle (of given density) the greater the terminal velocity. For most cloud droplets it varies directly as the square of the droplet radius: for raindrops the terminal velocity increases at a slower rate (see problem 4.1). For cloud droplets of radii 10 and 20 μm the values are about 1 and 5 cm s^{-1}, respectively, in still air†. Thus the droplets in a typical non-precipitating cloud remain suspended at a more or less constant level while a current of air with a velocity of a few centimetres per second flows up through them. The larger cloud particles, because of their higher speeds of fall, tend to separate out but evaporate very soon after they fall below the condensation level‡.

† The terminal velocity slowly decreases downwards as the particle encounters air of higher density.

‡ Drops have to be of at least 100 μm radius in order to reach the ground, even when the cloud base is low and the air below the cloud is moist.

4.1.4. *Ice nuclei*

It is possible, in the laboratory, to cool water so that it remains liquid at temperatures far below its nominal freezing point (0°C). The same is true of other liquids. This phenomenon, called supercooling, is seldom encountered to any great extent. However, cloud droplets in the atmosphere are very often supercooled and this has a number of important effects.

A cloud droplet freezes as soon as its temperature falls below −40°C. Between −40°C and 0°C it remains liquid unless it happens to contain, or encounter, an effective ice nucleus. Less is known, with reasonable certainty, about these particles than about the nuclei of condensation. It is thought, however, that they are much scarcer than condensation nuclei and are mainly particles raised from the Earth's surface. It has been shown in the laboratory that various types of naturally occurring particle, notably clays, cause (some) supercooled droplets to freeze below about −12°C; also, such particles have been identified in samples of ice clouds.

The ice nuclei, whatever they are, must be almost ineffective between 0°C and −12°C because nearly all the particles in clouds within this temperature range are water droplets. The proportion of ice crystals increases as the temperature falls below −12°C. Below about −30°C they begin to predominate: below −40°C all the particles are ice. The forms assumed by these crystals are very varied, e.g. ice needles, hexagonal plates, stars, etc.

4.1.5. *Ice-crystal clouds*

Ice clouds, such as cirrus, are much more tenuous than water clouds such as cumulus and stratus. There is also a big difference in the way in which they mix with the surrounding air. As previously mentioned, water droplets evaporate very quickly and so the edges of water clouds are relatively clear cut. On the other hand, ice crystals which fall from an ice cloud evaporate very slowly and frequently even grow and link up one with another. Thus the edges of ice clouds, and the spaces between individual parent clouds, are diffuse and indistinct. The reason is that an ice cloud first develops only when the volume of air which contains it is saturated with respect to water. This air is therefore supersaturated with respect to ice and so also may be the air surrounding the cloud out to some distance from it. The ice crystals eventually begin to evaporate when they reach air which is unsaturated with respect to them. The distinctive appearances of a water cloud and an ice cloud are well illustrated in Plate 1.

4.1.6. *Precipitation from water clouds*

From the information contained in § 4.1.3 it may be inferred that even the larger droplets of a typical water cloud would take 3 to 6 hours to fall from a height of 1 km†. So far as the production of rain is concerned it is therefore obvious that in a typical cloud the available water—usually about 1 g m^{-3}—is shared between far too many droplets. Concentration of the water into drops which are about 1 000 000 times fewer and 100 times bigger has two effects: the fall to Earth from 1 km height takes about 2 min; and these larger drops are, in any case, much less easily evaporated. Such drops, of radius about 1 mm, are typical rain drops. How, then, is the concentration into large drops brought about?

Calculation shows that the process that forms clouds (i.e. diffusion of slightly supersaturated water vapour to the droplets and condensation on them) is far too slow to account for their growth to the required size in a reasonable time. This is mainly because all the cloud droplets compete for the excess water vapour. Further, the droplets are so widely separated, on average, that their growth by random collisions is also very slow.

The collection of cloud water into raindrop size is, nevertheless, mainly by 'coalescence' resulting from collisions. However, the collisions are not random but arise because cloud droplets cover a range of sizes and correspondingly of terminal velocities (equally, of different rates of rise in a strong upcurrent). Thus small droplets which move upwards through the cloud as a whole overtake, and collide and unite with, larger droplets. When the resulting products reach a size such that they fall back through the cloud, they grow further when they overtake and coalesce with smaller droplets. It usually takes about an hour for raindrops to be formed in this way.

Strong upward motion, large depth of water cloud and high moisture content of the air are the factors which promote growth by coalescence. For these reasons, precipitation by this process (acting alone) is much more common in the tropics and sub-tropics than in higher latitudes‡.

4.1.7. *Precipitation from mixed clouds*

The other main method of cloud particle growth, which is important in all latitudes, involves the presence of (predominant) water droplets and ice crystals in the same ('mixed') cloud: it is called the Bergeron process. The temperature range involved is about $-12°C$ to $-30°C$. The air surrounding the particles is supersaturated with respect to the ice crystals; water vapour therefore condenses on them, the deficit being made good by the evaporation of water droplets (see §2.3.3). The ice crystals grow quickly at the expense of the more numerous droplets, and are soon large enough to fall through the cloud.

After this, coalescence takes over, the collisions being mainly between ice crystals and water droplets which freeze on impact. Aggregates of

---

† They of course evaporate long before they reach the Earth's surface.
‡ Precipitation in the form of drizzle, wherever it occurs, is nearly always of this type. (Drizzle comprises small rain drops of about 0·2 mm radius.)

44

many ice crystals, i.e. snowflakes, are thus formed and are infinite in their variety. If these fall below the 0°C level they begin to melt and arrive at the surface as rain or, if not completely melted, as sleet. Thus, most of the rain of middle and higher latitudes originates as snow.

Precipitation in the form of hail is a sure indication of very strong vertical currents within the cloud concerned (cumulonimbus) which often extends through nearly all the troposphere. Large hailstones have alternate rings of clear ice and frost. These rings are usually considered to show that the particles have been moved up and down repeatedly within the cloud, with alternations of growth conditions—in particular, of droplet size and degree of supercooling. When a hailstone collides with a supercooled droplet only a fraction of the droplet freezes instantly, the size of this fraction depending on the degree of supercooling (see problem 4.2). The large droplets flow round the hailstone on impact since their freezing (into clear ice) is not completed until the large amount of latent heat involved, which raises the temperature of the surface of the hailstone to 0°C, is lost to the surrounding air. In contrast, the freezing of small droplets is almost instantaneous: air is trapped between them and the deposit is one of opaque, frosty ice.

4.1.8. *Thunderstorm electricity*

Laboratory and outdoor experiments have confirmed that, within a cumulonimbus cloud, there is a wide range of possible mechanisms whereby electrical charges of opposite sign may become separated. Thus, for example, charges separate when large raindrops break up (as falling drops, of radius larger than about 3 mm, do); or when falling cloud droplets or ice crystals selectively capture atmospheric ions; or when ice particles of different temperatures collide; or when supercooled water droplets splinter on freezing. All these mechanisms, and perhaps others, probably operate within a cumulonimbus though with what relative importance is not known for certain.

Measurements show that positive charge accumulates near the top of the cumulonimbus cloud and negative charge near the bottom. When the gradient of charge between different parts of the cloud, or between neighbouring clouds, or between cloud and ground, reaches a critical value (many millions of volts per kilometre), the insulation is broken down by the passage of a very large current for a very short time. This is the lightning stroke and it occurs in a small minority of all cumulonimbus clouds. The violent heating and expansion of the air which simultaneously occur along the length of the stroke cause sound waves to radiate outwards, these being heard as thunder. Each 3-second interval between the seeing of the flash and the hearing of the thunder corresponds to a distance of 1 km between the discharge and the observer. Charge separation in the most active of thunderstorms is so rapid that lightning flashes may follow in quick succession for a considerable period of time.

The ionosphere acquires a net positive charge by the downward flow of negative ions from it to the positively charged tops of cumulonimbus clouds (which number perhaps a few hundreds of thousands per day over the Earth as a whole). The Earth acquires a negative charge because positive ions move upwards from higher points of it to the negative bases of overhead cumulonimbus ('brush discharge'); and also because negative charge is carried by lightning strokes to ground. These charges spread over the Earth and ionosphere and are prevented from accumulating by a small leakage current between them in fine-weather regions.

4.2. *Larger-scale processes*

Dew or hoar frost is deposited when the ground cools sufficiently by outward radiation. Rime and glazed frost occur when supercooled fog droplets and raindrops, respectively, contact surface objects whose temperature is less than 0°C. Nearly all fogs are caused by surface cooling of air, either by radiation or by transport across a colder surface, but a few are caused by rapid evaporation of water into the surface layer.

Away from the Earth's surface, cloud is formed when air cools by sufficient ascent: other processes are negligible in comparison. The ascent is, typically, either rapid but local, leading to heap (cumuliform) clouds; or it is slow but widespread, leading to layer (stratiform) clouds. Forced ascent of air over and downwind of high ground is often important. The great majority of clouds disperse without precipitating. Precipitation reaches the ground only when the ascent affects a deep enough layer.

4.2.1. *Surface cooling*

(1) *Dew and frost deposits*

On a clear night the Earth's surface loses heat by net radiation outwards. As we have seen, the temperature of a water surface is little affected, whether or not the water is calm. The same applies to a land surface if a strong wind is blowing because forced convection of the air then spreads the cooling through a deep layer.

If, on the contrary, there is little or no wind the land surface temperature falls quickly. When the air in contact with the ground reaches its dew point the water vapour which it contains begins to be deposited on the surface as dew. A 'radiation inversion' (of temperature) is formed at the surface. If the ground is moist the deposit is much increased by the diffusion of water vapour upwards in the soil and its condensation on the cold surface. A heavy downward deposition of dew from the air requires both moist surface air and also a light wind (no more than a few

46

knots) which, by causing gentle stirring of the surface layers, brings more water vapour down to the surface. Sometimes the dew later freezes on the ground; or, if the air is very cold and dry, the water vapour condenses directly as a fine ice deposit in a variety of picturesque crystalline forms: in either case the deposit is called hoar frost.

Very different in method of formation and appearance is the deposit of rough crystals, called rime, which is formed when supercooled fog droplets are carried by the wind against terrestial objects. Different again is the clear ice, called glazed frost, which is formed when rain or drizzle falls on a surface whose temperature is less than $0°C$†; on a road surface this is called black ice.

(2) *Radiation fog*‡

The rate of radiative cooling of a surface, and of the air near it, is much reduced by their absorption of the latent heat released when dew is deposited. The discontinuity in cooling rate is large when the temperature is high (see problem 4.3). If calm conditions persist throughout the night then cooling of the air which is not in immediate contact with the ground is dependent on the downward conduction of its heat to the ground. This process is not very effective, so that saturation and condensation are then confined to only a shallow surface layer, perhaps 1 m in depth. (Such 'ground fog' may temporarily thicken and deepen when post-dawn heating of the ground mixes up the surface layers.)

A light wind spreads the cooled air upwards and, if the surface layers are moist, condensation into fog is soon produced within them. When the fog has been formed the net radiative cooling to space is no longer from the ground but is from the top of the fog which is therefore marked by an inversion of temperature: the temperature within the fog is controlled mainly by radiative exchanges between the droplets and is therefore fairly uniform.

Radiation fog favours low-lying, marshy ground; also industrial regions where there are many hygroscopic nuclei§. It is common in valleys into which air, cooled by radiation, gravitates during the night. It tends to be most frequent and dense in autumn when there is plenty of moisture available, and in winter when the period of night cooling is long. Such fog in summer is uncommon and usually disperses quickly after dawn. In other seasons the weaker solar heating sometimes fails to raise the surface temperature sufficiently to disperse a thick fog during the day. In such a case the fog may persist for several days until perhaps the wind strengthens sufficiently to lift it into low stratus cloud.

† Drops of this size are not supercooled to any extent.
‡ By international definition, fog reduces surface visibility to less than 1 km.
§ The high concentration of smoke particles, mixed with water droplets, which builds up under a long-lasting surface inversion, is called smog.

(3) *Advection fog*

If surface air is cooled to its saturation point, not by outward radiation of the underlying ground, but by the air's transport across a cold surface then ' advection fog ' is formed: advection, in meteorology, means horizontal transport. Nearly all sea fogs are of this type. They occur when warm air moves across a cooler ocean (usually towards higher latitudes) and are able to persist because the heat which is extracted from the air is mixed downwards through the water and the temperature contrast between the sea and air at the surface is therefore maintained. When this contrast is very great (for example, when air from the Gulf of Mexico moves in spring and early summer across a sea current of Arctic origin near the North American coast) the loss of heat by contact is so rapid that the surface layers are foggy even when the wind is strong.

In the example quoted, the air is very moist and condenses onto the underlying sea surface. The situation is different in some other cases of advection fog. For example, the air which moves over the North Sea from the continent of Europe in spring and summer has a higher temperature but usually a lower dew point than the sea temperature. Water evaporates into the surface layer till its dew point is raised to the sea-surface temperature, its vapour pressure then being the same as that of the sea surface. At the same time the air is cooling by downward conduction. After a sufficient length of sea track the surface layers become saturated at the sea surface temperature and fog forms.

Coastal areas are often affected by the advection fog of adjacent seas. If the wind is strong enough to bring it far inland, it is also strong enough to lift it into low stratus by mixing with the drier air above. Fog rarely forms overland by advection alone but does occur in winter when moist sea air moves slowly inland, after a spell of cold weather. It is not long-lasting because equilibrium is soon established between the surface temperatures of the land and air.

4.2.2. *Evaporation*

(1) *Steam fog*

Paradoxically, a type of advection fog may also be formed when air, which is much *colder* than a water surface, is transported across it. The effect is seen in the steaming of exhaled breath; also, on a small meteorological scale, in the steaming of wet ground which is strongly heated by sunshine during cool, showery weather, or of lakes on clear autumn mornings when cold air from nearby hills flows across them. The phenomenon is usually called steam fog or arctic sea smoke—the latter because the large temperature difference which is required occurs over seas only in high latitudes (and there only locally).

The specified conditions favour continuous rapid evaporation because the vapour pressure at the water surface exceeds that of the air whether

48

or not the air is saturated. Air which is warmed and moistened at the surface is convected vigorously upwards and forms a supersaturated mixture, with resulting condensation into water droplets, in a layer the depth of which generally increases with the surface temperature contrast but which may be limited by a temperature inversion within the cold air.

(2) *Frontal fog*

Sometimes fog is formed by the evaporation of raindrops into cold surface air until condensation occurs in it. Since raindrops are not supercooled they would, in such circumstances, be falling from the warmer air above a frontal surface—see Chapter 8.

4.2.3. *Vertical motion*

(1) *Rapid local ascent*

The thermals which move upwards from a warm surface grow by adiabatic expansion and by mixing with the air through which they pass. Any which have enough buoyancy to reach their condensation level form small elements of cloud which, initially, evaporate almost as soon as they are formed. As they do so, they can be observed to descend slightly owing to cooling by the evaporation process. Since

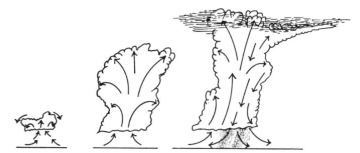

Fig. 4.1.   Stages in cumulus cloud development ('small' and 'large' cumulus and cumulonimbus).

the cloud elements, which are formed by following thermals, evaporate less readily into the moistened air there is a typical (unsteady) growth of clouds, horizontally and vertically, through the stages of 'small', 'medium' and 'large' cumulus to cumulonimbus (see fig. 4.1). The cloud development stops at any stage at which the supply of strong thermals ceases, often as the result of surface cooling in the late afternoon; or it stops when a very dry atmospheric layer is reached. On reaching a stable layer, the cloud top spreads horizontally. The typical anvil shape of the cumulonimbus top (see Plate 2) usually represents the bottom of the stable layer at the tropopause.

Clear signs of ice crystals near the cloud summit mark the cumulonimbus stage. Heavy, showery precipitation always occurs under the cloud at that time, sometimes with hail and thunder. Precipitation often starts at the earlier (large cumulus) stage, especially in low latitudes. A cumulonimbus is of the order of 5–10 km in width and depth. When the horizontal extent is much greater than this, the cloud mass comprises a number of 'cells' which have amalgamated. Updraughts in the cloud are often about 1 m s^{-1} but are sometimes as high as 20 m s^{-1} or more†. The compensating downward motion of the air which surrounds cumulus cloud is more widely spread and is therefore much gentler. Strong downdraughts, however, are induced in cumulonimbus by the falling precipitation. Aircraft flight within a large cumulonimbus cloud is made uncomfortable, perhaps dangerous, by these vertical currents.

Over land there is often a sun-facing slope or range of hills, or a patch of ground of low albedo, or other feature, which acts as a strong source of thermals when the ground is heated. Cumulus clouds which are continuously formed over such a feature move downwind from it and form a 'cumulus street' (Plate 3). This does not happen over the open sea and cumulus formation is much more cellular in character there.

Cumuliform cloud is not always formed by convection from the Earth's surface. A belt of cloud at a particular level may become unstable by lifting or by radiation exchanges. Cumuliform turrets then grow upwards from the cloud (its technical description is then castellanus) and may amalgamate into large cumulus or cumulonimbus.

(2) *Slow widespread ascent*

Most of the rainfall of tropical latitudes is from cumuliform clouds. In higher latitudes a greater percentage falls from stratiform clouds, i.e. clouds in layers of large horizontal extent. Precipitating stratiform clouds are formed by gradual upward air motion, of the order 0·1 to 0·2 m s^{-1}, over an extensive region and through a considerable depth: altostratus and nimbostratus are the main types involved. The precipitation is typically lighter, but longer lasting, than from cumuliform clouds. The upward motion which causes it is associated with fronts and other dynamical features (see Chapter 8).

(3) *Turbulence*

Widespread and persistent stratiform cloud is often formed in the upper part of a layer within which the flow is turbulent and there is a consequent vertical interchange of mass. The necessary conditions for this cloud formation are thorough stirring of sufficiently moist air through a restricted depth. (Without a restriction on depth cumuliform clouds would result.)

These conditions frequently occur within the 'friction layer', this being the lowest 0·5 to 1 km of the atmosphere, where turbulence is

† They may support very large hailstones which fall at this speed in still air.

induced by the air's passage over the Earth's surface. The degree of turbulence and the efficiency of mixing increase with the ground roughness and the wind speed. The effects of mechanical mixing within the friction layer are illustrated in fig. 4.2. Normal initial conditions of a lapse rate of temperature (but less than the DALR) and a lapse rate of mixing ratio are assumed. The temperatures of individual air particles change at the DALR: descending particles cause the temperature in the lower half of the layer to rise and ascending particles cause the temperature in the upper half to fall until the DALR is established throughout the layer and a temperature inversion is formed at its boundary with the unmixed air above. On the other hand, thorough mixing obviously leads to the same mixing ratio (the mean of the original distribution) throughout the layer. This value of mixing ratio often corresponds to saturation or supersaturation at the lower temperatures near the top of the layer. The water vapour then condenses to stratus cloud droplets above the ' mixing condensation level ', the lapse rate of temperature within the cloud being the SALR: see also fig. 5.4.

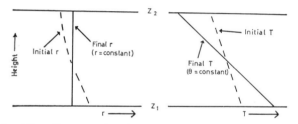

Fig. 4.2. The effect on a distribution of T and r of vertical mixing in a layer Z_1 to Z_2.

The above process explains why, when there is surface cooling, a strong wind usually leads to low stratus formation rather than fog; or leads to the lifting of fog, previously formed, into stratus. The same principle of cloud formation applies higher in the atmosphere on occasions when turbulence within a layer is confined by a temperature inversion or stable layer. Stratocumulus and altocumulus cloud sheets are often formed in this way. Sometimes the lower turbulence clouds are thick enough for light precipitation from them to reach the ground, but this is unusual.

(4) *Orographic effects*

Orography has important—often dominant—effects on the geographical distributions of clouds and precipitation. The influences are of various kinds but most of them result from the distortion of airflow over and near high ground.

A particular orographic effect not of this kind has already been mentioned—namely, the tendency for sun-facing slopes to act as strong

51

sources of thermals in sunny weather. This action is reinforced by the local flow of air upwards along a heated slope during the day (the 'anabatic wind ')†. For these reasons cumulus growth and thunderstorm development from daytime heating tend to be initiated over mountains rather than over adjacent plains and may be confined there throughout the day if the tropospheric winds are light. In some low and middle latitude regions this is a reliable feature of the climate.

Orographic cloud is most commonly formed on windward slopes: in the British Isles these clouds are most frequent in the west and southwest since the prevailing winds are from these directions. The surface air moving in from the sea is usually moist and requires little forced ascent to reach its ' lifting condensation level '‡. Often there is a complete c ver of low stratus cloud, formed by turbulent mixing, before ascent; the orographic effect is then to lower the cloud base (relative to sea level) by the amount which corresponds to the difference between the mixing and lifting condensation levels. Orographic stratus covers the windward slopes in ' hill fog ' and the cloud is often thick enough for persistent drizzle to fall from it.

If mountains bar the passage of a deep, moist current then the orographic cloud is thick and there are large amounts of ' orographic rain ' on the windward slopes. Approximate calculations of likely rainfall may sometimes be made in particular circumstances—although various assumptions are usually involved (see Chapter 5). The air is warmed and dried by its descent in the lee of high ground and rainfall is often absent there, or nearly so. When, in hilly regions, the prevailing winds are moist (as are the south-west to west winds of the British Isles) the effects of orographic ascent and descent on rainfall distribution and amount are very obvious on climatological mean charts. On such charts the area of low rainfall in the lee of a mountain range is sometimes called a rain shadow.

The warm wind often experienced in the lee of high ground is usually called a föhn, a name which originated in the Alps. If the surface air rises and then falls almost symmetrically in its passage over high ground, and if precipitation falls from it on the windward slopes, then its warming by descent (mostly at the DALR) exceeds its cooling by ascent (mostly at the SALR) and net warming is therefore to be expected—see fig. 5.5 (Chapter 5). This is no doubt at least a contributory factor in some occurrences of föhn. It is not, however, a necessary condition: there are occasions when appreciably warmer air exists to leeward than to windward yet the air is so dry that not even cloud is formed by ascent. The only explanation of such warming is that the air which reaches the

† Its converse (the ' katabatic wind ') is the gravitational flow down-hill on a clear night and involves a shallower layer. It has already been seen to have effects on the production of radiation fog and steam fog.

‡ The level at which the surface air, retaining its mixing ratio and cooling at the DALR during lifting, becomes saturated: see § 5.2.3.

ground in the lee of the hills was, on the upwind side of the barrier, at a level higher than the surface, i.e. it descends farther than it ascends.

Föhn conditions are often suitable for the formation of stationary air waves in the lee of high ground ('lee waves'). These have the form indicated in fig. 4.3 and are analogous to the stationary ripples which may be seen behind a submerged object in a shallow stream. The crests and troughs which form the individual waves remain stationary while the air blows through them. The wavelength (crest to crest) is usually about the same as the width of the obstruction but tends to increase with increasing height in the atmosphere. Conditions are most suitable for waves when a very stable layer of air (isothermal or inversion) is sandwiched between an unstable layer at the surface and another above. The waves often have their greatest amplitude (vertical development) within the stable layer.

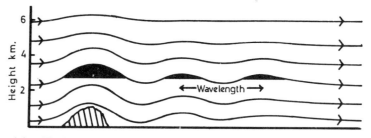

Fig. 4.3. Typical streamlines in lee wave formation. Lenticular cloud may be formed, their distance apart corresponding to the wavelength at the level concerned.

Lee waves, especially those of small amplitude, may exist without any visible indication of their presence. Glider pilots who are searching for lift then find it difficult to exploit the waves. At other times the air may contain just enough moisture for condensation to occur during the air's ascent to a wave crest, with subsequent evaporation during its descent to a wave trough. Smooth lens-shaped ('lenticular') clouds, as illustrated in Plates 4 and 5, are then formed—often at several levels in and above the stable layer, depending on moisture content—in a line at right angles to the hill. They remain almost stationary while the air moves through them: this is sometimes emphasized by the obvious motion of small cloud elements which may be formed by convection in an unstable surface layer.

Wave clouds commonly form at a height of 4 or 5 km (altocumulus) but they may be as low as 1 to 2 km (stratocumulus) or as high as cirrus levels (6 km to tropopause) or, very rarely may be within the stratosphere; their lenticular form and lack of significant motion indicate their method of formation. In favourable conditions they are usually conspicuous for a few hours after sunrise and before sunset but are often burned off, during the middle part of the day, by heat radiated and perhaps also convected from the ground.

Two further, uncommon, types of orographic cloud remain to be mentioned. A ' banner cloud ' may be formed when strong winds blow past an isolated, cone-shaped mountain (Plate 6). Pressure is reduced in the region immediately downwind of the peak; the air moves upwards, on the lee side of the mountain, into this region and the water vapour condenses to form a banner-shaped cloud fixed to, and downwind from, the mountain. Finally, there are ' rotor clouds ' which are very conspicuous with their saucer-like shape. This type of cloud is sometimes formed under a lee wave of large amplitude when an increase of wind in the vertical produces a large turbulent eddy, in the upper part of which water vapour condenses and under which the surface flow is usually reversed compared to elsewhere (see fig. 4.4).

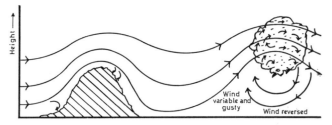

Fig. 4.4. Formation of turbulent rotor cloud: the axis of rotation is normal to the wind and usually below the cloud base.

4.3. *Cloud observations*

The acquired ability to ' read ' clouds intelligently is perhaps the most satisfying reward to be gained from a serious study of meteorology†. This art involves not only visual recognition of cloud types, but also reasoned assessments of their constitution, height of base and vertical extent, of how they have been formed and are likely to develop, and of wind velocities at various levels within and around them. Relevant information has been given in the preceding sections of this chapter. Here we summarize this and also add further points, some of which (for example, those relating to fronts, jet streams and wind shear) will be clarified in later chapters.

A striking general feature of clouded skies is the speed at which visible changes often occur in them and, to an even greater extent, in the individual clouds which comprise them. Even an apparently featureless layer of cloud is revealed by time-lapse photography to

† The following books will be found very helpful. *Cloud study: a pictorial guide.* F. H. Ludlam and R. S. Scorer. London (John Murray), 1957. 12s. 6d. *Cloud studies in colour.* R. S. Scorer and H. Wexler. Oxford (Pergamon), 1967. 42s.

change much more than would be judged from observation of it in real time.

4.3.1. *Cloud genera: their heights and composition*

The present classification of clouds, agreed and used internationally, stems from one originally made in the early 19th century, by Luke Howard whose remark " Like fishes inhabiting the bottom of an ocean, we are insensible to much of what passes over our heads " is probably even more apt now than when it was made. Howard recognized four principal cloud characteristics and used their Latin equivalents to describe them: they were cirrus (curl), cumulus (heap), stratus (layer) and nimbus (rain). These names are still used and have their original meanings: singly or in combination or, in two cases, in conjunction with the Latin prefix ' alto ' (high), they comprise the ten basic cloud genera.

The ten genera are stratus, cumulus, stratocumulus, cumulonimbus, nimbostratus, altostratus, altocumulus, cirrus, cirrostratus and cirro-cumulus. The first four of these are classed as ' low ' clouds with base below 2 km, altocumulus is a ' medium ' cloud with base between about 2 and 6 km, and the cirriform clouds are ' high ', with base above 6 km. Altostratus is usually at medium levels (sometimes higher), while nimbostratus occurs at medium and low levels. The cirriform clouds are mainly ice crystals; altostratus and the two precipitation clouds, nimbostratus and cumulonimbus, are mixed clouds; the others are mainly water clouds. The genera are illustrated in Plates 1 to 12.

4.3.2. *Cloud recognition and general features*

Cirriform clouds are easily recognized by their obvious large height and, in most cases, by their fibrous or milky appearance. Cirrus is usually in lines or bands from which ice crystals may fall and form long trails. (The ice-crystal anvil cloud at the top of a cumulonimbus is called false cirrus because of its special method of formation.) Cirrus clouds nearly always have a component of motion from the west. Where they occur in a well-marked line across the sky they usually coincide with a jet stream at that level. If the individual clouds are observed to move rapidly (with a speed of the order of 100 knots) in the general direction of the line in which they lie, this is confirmation of a jet stream.

Cirrostratus is a veil of ice crystals which, most typically, spreads across the sky from the west in advance of an approaching warm front. If the cloud is thin, a halo surrounds the sun or moon at an angular distance of $22°$ from it, the light being concentrated in a ring of this radius because of its refraction by the ice crystals. Sometimes cirrostratus and cirrus degenerate into a layer, or patches, of very small and closely packed cloud elements, often with a rippled appearance; this is cirrocumulus.

Altostratus is formed by slow, widespread ascent and is often in several layers. If it has been preceded by cirrostratus, the usual further progression is to nimbostratus from which persistent precipitation reaches the ground. The main cloud base may then be no more than a few hundred metres above ground, with thin patches of still lower stratus cloud below.

Cumulus clouds are initially detached, with fairly flat bases at a uniform level and with rounded tops. If they form early in the day in a sky which is of a deep blue colour, quick development into large cumulus, then into cumulonimbus and heavy showers, is to be expected. If, on the other hand, small cumulus clouds form during the forenoon in a light blue, hazy sky† the clouds are much more likely to disperse as the day progresses; this is because the thermals from the warmer surface later reach still drier levels of the inversion layer. However, if the surface layers are unusually moist, the clouds tend to spread sideways at the inversion and so form stratocumulus which may persist and reduce considerably the amount of sunshine at the ground.

Cumulonimbus, with its great vertical extent and glaciated top, is easily recognized at a distance, but much less easily if it is overhead, especially if intermingled with layer clouds. Rapid changes of precipitation intensity strongly indicate that the overhead cloud is cumulonimbus; the occurrence of either hail or thunder is a sure sign that this is so.

Stratus is formed by orographic ascent or by turbulent mixing at low levels. An almost featureless layer formed in the latter way often changes into a stratocumulus layer, with variations in appearance from barely perceptible regularities of light and shade to cloud masses and clear spaces of about equal size. The change is caused by the development of cellular up-and-down motion within the cloud when it is made unstable by radiative cooling at the top. The same process changes altostratus into altocumulus, with the result that these cloud types are often seen together.

4.3.3. *Effects of vertical wind shear*

Except for cloud formed in lee-wave conditions, the wind direction at cloud level is easily inferred by observing its motion relative to a suitable foreground as frame of reference (§7.4.4). The relative wind speeds at different levels may be assessed if it is borne in mind that, for a given angular velocity, the linear velocities of clouds are proportional to their heights—see fig. 4.5. An accurate assessment depends on a reliable estimate of relative cloud heights, for which experience is needed.

The ' shear vector ' which represents the change of wind velocity with height is the vector difference between the wind at a higher level and that at a lower level. If, for example, westerly winds increase with

† Indicating a dry inversion layer at 1 to 2 km height.

height through a layer then the shear vector is westerly: more generally, a change of direction is also involved. This vector is often approximately indicated by certain cloud characteristics, such as the slant of a trail of ice crystals falling from high or medium cloud, or the slant of a cumulus tower (the upwards slant is in the direction of the shear vector in each case), or the orientation of a cumulonimbus anvil.

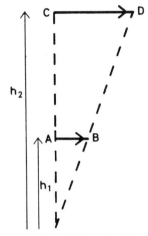

Fig. 4.5. For a given angular velocity the linear velocity of clouds at heights h_1 and h_2 is proportional to their height.

Wind shear in the vertical is itself a common cause of cloud formation. If, say, a layer of air is moving faster than the layer underneath they interact at their boundary so as to reduce the difference in their speeds. Turbulent mixing occurs through a limited depth and if the air is moist enough cloud is formed. Jet stream cirrus is an example which is sometimes very restricted laterally. More generally, a layer of cloud covering all or most of the sky is formed and may be at any level. If the shear vector is strong then the clouds are arranged in 'rollers' or 'billows' which run at right angles to the shear vector and which usually move in the direction of the shear vector.

4.3.4. *Cloud classification for forecasting*

Even if the above account were much more complete than it is, it would not provide a ready explanation for every single cloud form, or cloud combination or characteristic that may be observed. It is, in fact, part of the fascination of clouds that this is so. Nevertheless, application of these main principles can provide a logical explanation of most states of the sky.

The international cloud codes which are used for professional weather-forecasting purposes provide for nine cloud types, or combinations of

57

types, at each of three different levels. There is no basic departure from the ten cloud genera (and also the species castellanus which has special significance) but the extra groups enable distinctions to be made between different stages of cloud growth or methods of formation, where applicable, and also allow for various combinations of clouds at a given level. The emphasis is on general states of sky and on the methods or stages of cloud development.

Questions on Chapter 4

4.1. It is found that the drag exerted on small cloud droplets during their fall is proportional to the velocity and radius: show that their terminal velocity is proportional to the square of the radius. For large raindrops the drag is proportional to the squares of velocity and radius: show that, in this case, the terminal velocity is proportional to the square root of the radius.

4.2. What fraction of a supercooled water droplet at a temperature of $-8°C$ freezes instantly when it collides with a solid object? The specific heat of water $= 4\cdot2 \times 10^3$ J kg^{-1} K^{-1} and its latent heat of fusion $= 3\cdot35 \times 10^5$ J kg^{-1}.

4.3. Find the temperature above which saturated air which is cooled by its contact with an underlying surface loses water more readily than it loses the associated heat. (Use a tephigram illustrated in Chapter 5 to determine dr_w/dT.)

the tephigram

THE meteorologist is faced with the problem of how to represent the state of the atmosphere in a form best suited for analysis. When the range of possibilities of variation of the meteorological elements, in three dimensions, is considered it is clear that no single form of representation can describe the system adequately. A ' synoptic chart ' which shows the distribution of selected elements, over a geographical region at a particular time and specified level, is one form of representation, of which the surface synoptic chart (weather map) is the most familiar example. In this chapter we are concerned with almost the antithesis of this, namely with vertical variations at a particular place and time. These conditions are not strictly observed in routine observations because the conventional sounding balloon, carrying its instrument load, takes about an hour to reach its normal ceiling and is then a considerable distance (typically 50 to 100 km) from its launching point. No allowance is made for this discrepancy, the sounding being regarded as representing the conditions in the vertical at the launching site at about the mid-time of flight.

The standard elements observed in a sounding are those of pressure, temperature and humidity: heights and pressure levels are related by the altimeter equation (equation (2.10) or (2.11)). These elements may be very simply represented on a graph which, for example, has linear scales of temperature and humidity as abscissae and height as ordinate, pressure then being on an approximately logarithmic scale decreasing upwards. A simple graph of this kind is adequate for certain purposes but not when accurate calculations of pressure variation with height, or of energy gains or losses in various processes, are involved†.

The tephigram is one of a number of diagrams which, though at first sight rather complicated, avoid these difficulties. The apparent complexity is caused by the inscription, on the diagrams, of five main sets of reference lines, namely isobars, isotherms, lines of saturation moisture content, and lines representing the dry adiabatic and the saturated adiabatic lapse rates of temperature. These lines are needed both for plotting the observations and as aids in their analysis. The actual construction, from first principles, of a diagram of this type is a

---

† Because equal area at different points on such a graph represents different amounts of energy: see §§ 5.1.1 and 5.5.1.

useful exercise for its better understanding. The method of construction is therefore explained for the tephigram (the same principles may be applied to other, similar adiabatic diagrams) before some of its applications are illustrated.

5.1. *Construction of the diagram*

5.1.1. *Coordinates: area and energy*

The tephigram (or $T–\phi$ gram) is so called because its co-ordinates are temperature (T) and entropy (ϕ)†. Entropy is a concept with which we shall not be concerned, except to show its relationship to potential temperature (θ).

By definition, the change of entropy ($d\phi$), per unit mass of a substance which undergoes a reversible thermodynamic process, is related to the amount of heat (dq) which it absorbs at temperature T by the equation

$$d\phi = \frac{dq}{T}.$$

$$\text{So } d\phi = \frac{c_p dT - v dp}{T} \quad \text{from equation (3.7)}$$

$$= \frac{c_p dT}{T} - \frac{R dp}{p} \quad \text{from the gas equation;}$$

$$\therefore \quad \phi = c_p \ln T - R \ln p + \text{constant.} \tag{5.1}$$

The potential temperature (θ) is, from equation (3.9), given by

$$\frac{\theta}{T} = \left(\frac{1000}{p}\right)^{0.288} = \left(\frac{1000}{p}\right)^{R/c_p};$$

$$\therefore \quad \ln \theta - \ln T = R/c_p \ln 1000 - R/c_p \ln p;$$

$$\therefore \quad c_p \ln \theta = c_p \ln T - R \ln p + \text{constant.} \tag{5.2}$$

Thus, from equations (5.1) and (5.2),

$$\phi = c_p \ln \theta + \text{constant.} \tag{5.3}$$

A linear scale of ϕ is, therefore, a logarithmic scale of θ (in K). The ordinate scale of the tephigram is log θ, calculated for a suitable range of θ in K though the labelling is in °C‡: the abscissa is a linear scale of temperature (see fig. 5.1 *a* which represents only a portion of the diagram). The horizontal lines are dry adiabatics (since $\theta =$ constant along them) and the vertical lines are isotherms ($T =$ constant).

It may be shown that unit area (A) on the tephigram represents at all

---

† Now usually designated S.
‡ There is a convention here which must be understood clearly. In thermodynamic equations T always stands for absolute temperature in K; however, in graphs and tables T may represent the corresponding Celsius value, for practical reasons.

points a definite amount of energy, as follows. From equation (5.2),

$$c_p \ln \theta = c_p \ln T - R \ln p + \text{constant};$$

$$\therefore \quad c_p \frac{d\theta}{\theta} = c_p \frac{dT}{T} - R \frac{dp}{p};$$

$$\therefore \quad c_p \, d(\ln \theta) = c_p \frac{dT}{T} - R \frac{dp}{p};$$

$$\therefore \quad c_p T \, d(\ln \theta) = c_p dT - RT \frac{dp}{p}.$$

If $c_p T \, d(\ln \theta)$ is integrated round a closed curve on the diagram then

$$c_p \oint T \, d(\ln \theta) = c_p \oint dT - R \oint \frac{Tdp}{p}$$

$$= -R \oint \frac{Tdp}{p} \quad \left(\text{since } \oint dT = 0\right)$$

$$= -\oint v dp, \quad \text{from gas equation,}$$

$$= \oint p dv - \oint R dT, \quad \text{from equation (3.4),}$$

$$= \oint p dv;$$

i.e. $dA = dW$, see equation (3.5).

5.1.2. *Isobars*

From the definition of θ, the 1000 mb line obviously joins those points where $T = \theta$, as shown in fig. 5.1 a. Other selected isobars are drawn (on the actual diagram at 10 mb intervals) by the following procedure, the 900 mb isobar being quoted as an example:

(a) Find θ/T for $p = 900$ mb from equation (3.9).
(b) For this value of θ/T find θ for $T = 0°C$ (273 K), $-10°C$, etc.
(c) Plot the points $\theta_1 T_1$, $\theta_2 T_2$, etc.
(d) Label the curve drawn through these points 900 mb.

5.1.3. *Saturation mixing ratio* (r_w) *lines*

The construction of these lines is based on equation (2.20), in the form

$$r_w = \frac{5}{8} \frac{e_w}{p}.$$

61

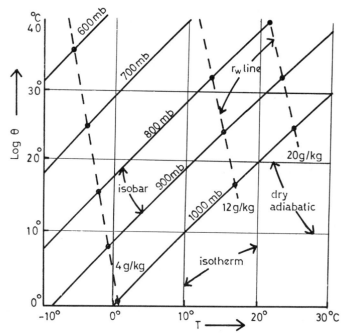

Fig. 5.1 a. Construction of isobars and saturation mixing ratio lines on a tephigram.

The method of construction, with $r_w = 20$ g kg^{-1} taken as an example, is as follows:

(a) Find e_w/p for $r_w = 20$ g kg^{-1}.
(b) For this value of e_w/p find e_w for $p = 1000, 900, 800$ mb etc.
(c) Find T corresponding to each e_w ($e_w = f(T)$ only).
(d) Plot the points $p_1 T_1$, $p_2 T_2$, etc.
(e) Label the curve through these points 20 g kg^{-1}.

The lines representing 4, 12 and 20 g kg^{-1} are illustrated in fig. 5.1 a.

5.1.4. Saturated adiabatics

During saturated ascent of air, adiabatic cooling on the one hand and warming by absorption of latent heat of condensation on the other, occur simultaneously. The saturated ascent may be assumed to be approximated by the condensation of a known amount of water vapour (Δr) at constant pressure, followed by the ascent of the air (which is now unsaturated, having lost water vapour and having been warmed by latent heat absorption) at the DALR till it becomes saturated†. The temperature rise ΔT is found by equating the heat absorbed by the air

† This assumption is obviously justified when Δr is infinitesimal.

to that released by condensation of the water vapour, i.e.

$$c_p \Delta T = L \Delta r,$$

where c_p is the specific heat at constant pressure $(10^3 \text{ J kg}^{-1}\text{K}^{-1})$ and L the latent heat of condensation of water $(2\cdot5 \times 10^6 \text{ J kg}^{-1})$: ΔT is thus $2\cdot5$ K per gramme of water vapour condensed from a kilogramme of air, and is directly proportional to the mass of water vapour condensed.

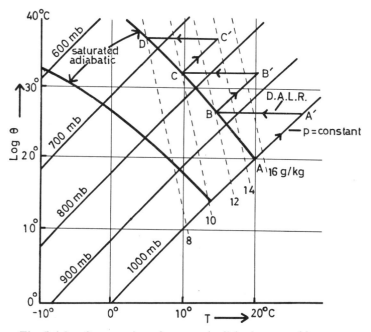

Fig. 5.1 b.   Construction of saturated adiabatics on tephigram.

In fig. 5.1 b, saturated air at A (1000 mb, 20°C) is assumed, by interpolation between the r_w lines, to have a mixing ratio of 15 g kg⁻¹. If 3 g water vapour is the amount selected to be condensed before ascent, the air warms to 27·5°C while still at 1000 mb (A′).   It now holds 12 g kg⁻¹ and, on ascent, cools at the DALR till it reaches saturation on intersecting the $r_w = 12$ g kg⁻¹ line at B.   The straight line AB is a good approximation to the (curved) saturated adiabatic AB.   The further path curves BB′C and CC′D are both based on condensing 2 g water vapour $(\Delta T = 5$ K) before ascent.   The smooth curve ABCD is the saturated adiabatic.   The complete tephigram has those curves drawn at selected intervals.   Their slope decreases with temperature owing to the smaller amount of water vapour contained in saturated air at low temperatures.

63

5.1.5. *Height variation*

In its printed (British Meteorological Office) form the tephigram is rotated by about 45° clockwise relative to fig. 5.1 *b*: pressure and height then vary almost along the vertical, the former logarithmically to about 200 mb and the latter linearly to about 12 km. Heights in a 'standard atmosphere' (mean sea-level pressure 1013·25 mb and temperature 15°C, lapse rate 6·5 K km^{-1} to 11 km, then isothermal) are printed at 50 mb intervals on the diagram; while thicknesses of 10 mb layers at all temperatures and pressures are shown at 100 mb intervals[†].

5.2. *Simple graphical computations*

The results of a vertical sounding are issued as values of temperature (T) and dew point (T_d) at all 'significant' pressure levels (p). Each corresponding value of p, T and of p, T_d is a unique point and is marked on the diagram: the successive p, T points are joined by a solid line (the temperature environment curve) and successive p, T_d points by a pecked line (the dew-point environment curve).

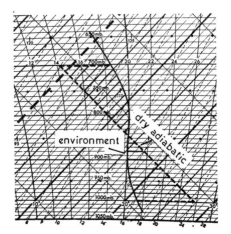

Fig. 5.2 *a*. Height determination by dry adiabatic method:
$Z_{700}-Z_{1000} = 29 \cdot 0/9 \cdot 8 = 2960$ m.

5.2.1. *Height*

In a particular sounding, the height of a given pressure level above the surface or above mean sea level, or the thickness of an isobaric layer, is obtained by the dry adiabatic method or by the isothermal method. The former, illustrated in fig. 5.2 *a* for the 1000 to 700 mb layer, consists of drawing that particular dry adiabatic which equalizes the

---

† The 'MINTRA' line on the diagram indicates the temperature (a function of pressure) above which exhaust trails from aircraft are unlikely to form: it corresponds very closely to $r_w = 0 \cdot 55$ g kg^{-1}.

areas indicated (i.e. between pressure level, dry adiabatic and T environment curve), reading off the temperature difference (ΔT) along the dry adiabatic between the two pressure levels, and dividing ΔT by 9·8 to obtain the height interval in km†. The isothermal method (fig. 5.2 b) involves selecting the particular isothermal which equalizes the areas between pressure level, isothermal and T environment curve and then either using the 10 mb values of thickness printed on the form, or substituting the isothermal value T in the altimeter equation (equation (2.11)). Strictly, with either the dry adiabatic or the isothermal method, a virtual temperature environment curve should be used—see §2.3.2.

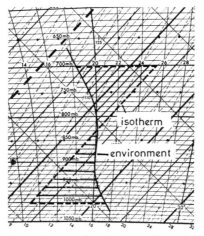

Fig. 5.2 b. Height determination by isothermal method: $Z_{700} - Z_{1000} = (883 + 990 + 1120)$ m $= 2993$ m by interpolation from printed figures for 10 mb thicknesses.

5.2.2. Humidity elements

These are easily obtained from the tephigram, as shown by the following examples which are illustrated in fig. 5.3 for the air sample $p = 750$ mb, $T = 8°C$, $T_d = 3°C$.

Mixing ratio (r): Since for an air sample, $r = r_w$ at T_d, then r is given by the value of r_w interpolated at T_d, i.e. $r = 6·3$ g kg^{-1}. It follows also that the T_d environment curve is also the r environment curve.

Relative humidity (U): Since

$$U = \frac{r}{r_w} \times 100\%, \text{ then } U = \frac{r_w \text{ at } T_d}{r_w \text{ at } T} \times 100\%,$$

i.e.

$$U = \frac{6·3}{9·0} \times 100\% = 70\%.$$

† DALR (Γ) $= -9.8$ K km^{-1}.

65

Vapour pressure (e): From equation (2.20),

$$e = \frac{8}{5}rp = \frac{(r_w \text{ at } T_d)p}{0.62}$$

$$= 7.6 \text{ mb.}$$

The arithmetic is avoided if it is noted that $e = e_w$ at T_d and that $e_w = f(T)$ only. Thus we may obtain e by using any pair of corresponding values of r and p along the isotherm T_d. If, then, we proceed

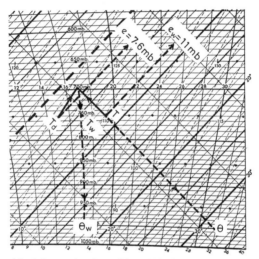

Fig. 5.3. Graphical determination of humidity parameters from p, T, T_d plot.

up or down along the isotherm from a given T_d to the particular pressure level 620 mb the interpolated value of $r_w (\text{g kg}^{-1})$ at this level is, numerically, the vapour pressure (e) in millibars. Thus from fig. 5.3, $e = 7.6$ mb. (The student may confirm the constancy of the product $r_w p$ along an isotherm.) The saturation vapour pressure is similarly obtained by proceeding to 620 mb along the T isotherm; thus $e_w = 11$ mb.

Wet-bulb temperatures (T_w and θ_w): A useful theorem relating to T_w is that of Normand, namely, "for an unsaturated sample of air, the dry adiabatic through T, the saturation mixing ratio line through T_d and the saturated adiabatic through T_w meet in a point". A corollary is that if, having determined the meeting point of the first two lines, we return from it along the saturated adiabatic to the original pressure level, the temperature attained there is T_w. Thus, in fig. 5.3, $T_w = 5.3°C$. If motion along the saturated adiabatic from T_w is continued to the 1000 mb level, the temperature attained is the wet-bulb potential temperature (θ_w)—i.e. 17°C. (This is analogous to potential temperature which is attained by dry adiabatic motion from T to 1000 mb; here $\theta = 32.3°C$.)

66

5.2.3. *Condensation levels*

The lifting condensation level of a mass of air is the (pressure) level at which the r_w line through T_d meets the dry adiabatic through T; this level is 695 mb in fig. 5.3. (T_d of an individual mass changes during unsaturated vertical motion along the r_w = constant line because the mixing ratio of the mass is unchanged.)

The mixing condensation level for a specified layer is found by determining the level at which the average values of θ and r for the layer intersect, these averages being best determined by equalizing areas as indicated in fig. 5.4. The convective condensation level is considered in § 5.5.1.

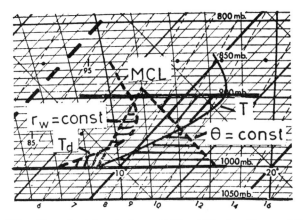

Fig. 5.4. The effect of mixing the layer 1000 to 900 mb without adding heat or moisture. The T curve changes to θ = const (areas above and below equal) and the T_d curve changes to r = const (areas right and left of environment curve equal). The MCL is at 890 mb.

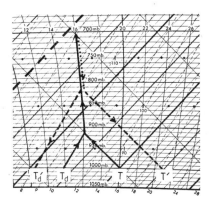

Fig. 5.5. Illustration of föhn effect: ascent curve in full, descent curve dashed.

5.2.4. *Föhn effect*

A simple example of the föhn effect (Chapter 4) is illustrated in fig. 5.5. The assumptions are that surface air, containing 10 g kg^{-1} water vapour at 1000 mb, rises in orographic ascent to 700 mb and subsequently descends to the 1000 mb level on the lee side; it is further assumed that 2 g of water vapour is precipitated per kilogramme of air during the ascent. The lifting condensation level of the surface air is 950 mb: cloud is formed at this level and the subsequent ascent (to 700 mb) is at the SALR. Descent of the cloudy air is initially at the SALR but increases to the DALR beyond the $r_w = 8$ g kg^{-1} level. T therefore rises, on descent, to $T' = 25°C$ at the surface (20°C before ascent); while T_d rises along the $r_w = 8$ g kg^{-1} line to $T_d' = 10·5°C$ at the surface (14°C before ascent).

5.3. *Precipitable water and precipitation rate*

5.3.1. *Formula and calculation*

The mass of water vapour contained in a vertical column of air, of unit cross-section, between any two levels (e.g. the Earth's surface and the top of the atmosphere) is easily determined from a tephigram. The formula which is applied is derived as follows.

In a small element of moist air, of thickness dz and of unit cross-sectional area, the mass of water vapour contained (density ρ_w) is

$$dm_w = \rho_w dz.$$

Thus, in a column extending from heights z_1 to z_2, bounded by pressure levels p_1 and p_2,

$$m_w = \int_{z_1}^{z_2} \rho_w \, dz$$

$$= - \int_{p_1}^{p_2} \frac{\rho_w dp}{\rho g} \quad \text{(from hydrostatic equation)}$$

$$= \frac{1}{g} \int_{p_2}^{p_1} r dp$$

$$= \frac{\bar{r}(p_1 - p_2)}{g}. \tag{5.4}$$

where $\bar{r}$ is the average value of r for the layer p_1 to p_2. For the normal units of r in g kg^{-1}, $(p_1 - p_2)$ in mb and g in ms^{-2}, m_w is given in units of 10^{-1} kg m^{-2}; this corresponds to a water depth in units of 10^{-1} mm.

m_w is called the precipitable water† and usually refers to the whole atmospheric column. It is found by totalling the contributions of selected successive layers; the appropriate $\bar{r}$ in a layer is obtained by

---

† m_w is related only in a general way to precipitation rates and amounts, other important factors being stability and vertical motion.

equalizing areas as in finding the lifting condensation level. Thus, referring to fig. 5.5, m_w for a saturated column of lapse rate given by the curve $\theta_W = 20°C†$ (1000 to 700 mb only, in steps of 100 mb) is

$$m_w = \frac{1}{g} \int_{700}^{1000} \bar{r} \, dp$$

$$= \frac{[(13\cdot9 \times 100) + (11\cdot8 \times 100) + (9\cdot7 \times 100)] \times 10^{-1}}{9\cdot8} \text{ mm}$$

$$= 36\cdot1 \text{ mm}.$$

5.3.2. *Precipitation rate*

An extension of this method yields estimates of precipitation rates, on various assumptions. Consider again saturated air with a lapse rate which corresponds to $\theta_W = 20°C$. If the layer with base at 1000 mb and thickness 100 mb ascends at the SALR till its base reaches 950 mb, then the initial $\bar{r}$ for the layer is $13\cdot9$ g kg^{-1} and the final $\bar{r}$ is $12\cdot7$ g kg^{-1}. If all the excess water vapour is assumed to be precipitated, its depth is $\Delta m_w = (13\cdot9 - 12\cdot7) \times 100/98$ mm $= 1\cdot2$ mm. If the ascent of 50 mb is completed in, say, 2 hours (upward velocity about 6 cm s^{-1}) this layer contributes $0\cdot6$ mm hr^{-1} to the precipitation. More generally, if a layer of thickness δp is lifted in a time δt by an amount such that the initial and final values for the layer are r_0 and r_1, then the precipitation rate is

$$\frac{\delta P}{\delta t} = \frac{r_0 - r_1}{g} \frac{\delta p}{\delta t}. \tag{5.5}$$

If the base of the layer ascends by δz_b, then

$$\frac{\delta P}{\delta t} = \frac{r_0 - r_1}{g} \frac{\delta p}{\delta z_b} \frac{\delta z_b}{\delta t}$$

$$= \rho_0 w_0 (r_0 - r_1), \quad \text{from equation (2.7)}, \tag{5.6}$$

where ρ_0 is the density and w_0 the vertical velocity of the base of the layer concerned. In the application of equation (5.5) or (5.6) to obtain an estimate of orographic rainfall, say, assumptions have to be made concerning the magnitude and rate of ascent of air at different levels—see, for example, problem 5.6.

5.3.3. *Water content of convective clouds*

The water content of clouds which are formed by convection from the surface is easily estimated from the tephigram on the assumptions that all the water which is condensed by adiabatic expansion is retained

† θ_w values are indicated at 700 mb.

‡ Neglect of the water content of clouds is justified if cloud and precipitation are persistent.

within the rising air and that the cloud does not mix with its surroundings. The calculated water content (g m^{-3}) increases with the temperature at the condensation level and also with height within the cloud. (Values obtained in this (adiabatic) way are maximum values; observed values are less because effects of mixing of drier surrounding air into the cloud updraughts and turrets are spread through the cloud.)

5.4. The effects of vertical motion on lapse rate

5.4.1. Unsaturated or saturated motion

The lapse rate and stability of a layer of air are, in general, changed if it moves upwards or downwards. We recall first, from Chapter 3, that an increase of environmental lapse rate implies a (relative) increase of instability. Consider now, in fig. 5.6, the isothermal layer AB. Assume unsaturated adiabatic ascent of this layer by 100 mb to A_1B_1

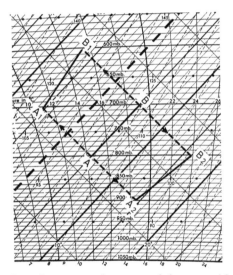

Fig. 5.6. The effect of unsaturated ascent and descent, without convergence or divergence, on an isothermal layer AB.

in such a way that the cross-sectional area of the layer is unchanged: the column stretches vertically because of the reduced average pressure on it but the pressure difference between the top and bottom of the layer is unchanged because the mass of air is the same as in AB. In fig. 5.6 it is seen that the top of the layer has cooled further than the base owing to the vertical stretching and that this has resulted in a change from isothermal to lapse conditions, i.e. a relative decrease in stability. If other examples are considered it is found that unsaturated ascent of a layer causes its lapse rate to approach the DALR; also that this is true whether the original lapse rate is less than the DALR, as in the example considered, or greater than the DALR†. It may similarly be verified that saturated ascent of a layer tends to establish the SALR within it, whatever the initial lapse rate.

The descent (subsidence) of the isothermal layer from level AB to level A_2B_2 is seen in fig. 5.6 to establish an inversion within it. More generally, large-scale descent of air causes its lapse rate to depart further from the DALR; in practice, since the original lapse rate hardly ever exceeds the DALR, descent causes the air to become more stable.

The illustrations of fig. 5.6 are based on a constant cross-sectional area of column and therefore a constant pressure difference between base and top. The actual effects are, in fact, accentuated because there is a tendency for a subsiding layer to spread sideways (i.e. to diverge horizontally) and so to shrink vertically; conversely, for horizontal convergence to be directly linked with vertical ascent and stretching of a layer of air. These are the typical motions within the anticyclone and the depression, respectively—see §8.7.1.

5.4.2. Potential (convective) instability

For a particular vertical distribution of humidity, forced ascent of a stable layer causes it to be become unstable, as follows. The layer AB in fig. 5.7 has a lapse rate less than γ_s and so is stable in the sense that it resists the motion of air parcels (saturated or not) through it. Suppose, however, that the layer is lifted bodily—for example by ascent over high ground—in such a way that the same 'pressure depth' of layer (100 mb) is maintained. With the distribution of humidity shown, the base of the layer becomes saturated on ascent sooner than does the

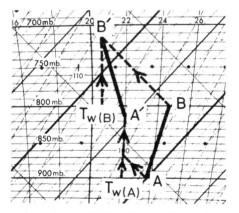

Fig. 5.7. Illustration of potential instability. Layer AB is made unstable by lifting to saturation.

† This condition scarcely arises in the free atmosphere.

top of the layer (note that wet-bulb temperatures are plotted in this instance). The fraction of the ascent at the DALR (Γ) is therefore greater for the top of the layer than for the base and the lapse rate within the layer steepens. Since each part of it ascends, after saturation, along its original θ_W line the lapse rate in the layer after the whole of it is saturated exceeds the SALR (γ_s) if $\Delta\theta_W/\Delta z < 0$ in the original distribution (i.e. if the original lapse rate of T_W exceeds γ_s). This is the criterion for the existence of ' potential instability '†, so called because, while it is often present within a particular layer or layers of a sounding (identified simply by a positive lapse of θ_W), it requires for its release a specific minimum (but often large) forced ascent of the layer which, much more often than not, is not forthcoming. Potential instability may also be released in a layer which becomes saturated at T_W by the evaporation of falling raindrops within it. When potential instability is released, cumuliform rather than stratiform clouds result and there is heavy rain.

5.5. *Tephigram analysis*

5.5.1. *Latent instability*

The temperature-environment curve of fig. 5.8 is typical of a conditionally unstable air mass ($\gamma_s < \gamma < \Gamma$) in which a surface inversion has formed overnight by net radiative cooling, in conditions of clear sky and light wind. We examine the likelihood that cumulus clouds, and perhaps showers, will develop within this air mass during the day.

Solar heating of the ground leads to vertical mixing of the air in a layer, the depth of which gradually increases as the surface temperature rises. The DALR is established in this layer (e.g. BC in fig. 5.8) and the environment curve is then BCDEJ. If forced ascent (orographic lifting, say) were to carry the air upwards from C it would become saturated at G, this being the lifting condensation level (LCL) corresponding to the surface value of r ($9 \cdot 5$ g kg^{-1}) which is expected to be reached by evaporation and mixing. Beyond G the forced ascent would be at the SALR (path GE) and the area CGEDC between the lifting path curve and environment is a measure of the energy required for lifting the air. E is the ' level of free convection ' (LFC) above which the air parcel is buoyant and accelerates upwards; area EFJE is a measure of the energy *gained* by the parcel in this part of the ascent. This type of instability with respect to parcel ascent is appropriately called latent, the necessary and sufficient condition for which is that a saturated adiabatic through some point on the T_W environment curve should intersect the T curve at a higher level. Extensive thundery outbreaks are often associated with instability of this type at perhaps 2 to 5 km height. (In practice, it is often difficult to know whether

---

† Also termed convective instability.

it is latent or potential instability that is released because the existence of the former necessarily implies the latter; the converse is not necessarily true, though it usually is.)

As surface temperatures rise above B, it is readily confirmed from fig. 5.8 that the corresponding LCL (point G) rises, while the LFC (point E) falls. When points G and E coincide (at D) the corresponding surface temperature is the minimum for which cloud formed by surface convection will rise by buoyancy. HDK is the path curve, the base and top of the cumuliform cloud thus formed being at D and K

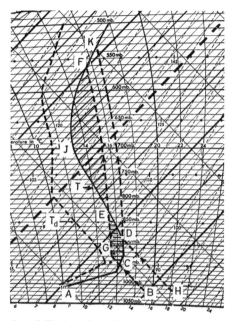

Fig. 5.8. T (——) and T_d (– – –) environment curves. Arrowed path curves shown from B and H. Negative ($\equiv$) and positive (////) areas represent amounts of energy supplied to, and acquired by, the air in its ascent above C.

respectively; area DKFJED represents the energy acquired in ascent. D is the convective condensation level (CCL) for the environment at this particular stage. (More generally, the CCL lies to the right of the night environment curve and rises with increase of surface temperature, as the bases of cumulus clouds are also observed to do†.)

The problems that arise in predicting from a profile such as fig. 5.8 are whether forced ascent of the required magnitude (which itself varies with surface temperature) will release the latent instability; and, if not

† A useful rule of thumb is that cumulus cloud base = $125(T - T_d)$ metres where T, T_d in °C refer to the surface.

whether the surface temperature will rise to the value (H) required for cloud formation by free convection: in either case the appropriate surface value of r has to be estimated. The parcel method, simple though it is, is generally reliable because the main complicating factors tend to counteract each other, as follows. The lapse rate of cloud within which dry environmental air has been entrained exceeds the SALR assumed in parcel ascent, and the real buoyancy of the cloud is less on this account. On the other hand a dry environment has a relatively high density, and cloud buoyancy is thereby increased (§3.2.5).

5.5.2. *Air mass characteristics*

The production, properties and transformation of air masses, and their recognition from surface observations, are discussed in Chapter 8. Diagrams such as the tephigram afford a more basic approach to air mass recognition and classification because they provide information on the vertical variations of temperature and humidity and because, at higher levels, these properties are not influenced by local (land surface) effects.

(1) *Conservative properties*

Various processes nevertheless operate also at higher levels to change air mass temperature and humidity. These include vertical motion, and also condensation of water from the air or evaporation of rain into it. Non-adiabatic heating or cooling of the air by radiative transfer is unimportant in the short term—one or two days, say—but increases in importance as time progresses. In order to trace and identify air masses attention is focused on certain properties which remain (relatively) unchanged during the various processes to which the air may be subjected: such properties are 'conservative' with respect to the specified processes.

Air temperature (T) changes at the DALR on unsaturated ascent or descent and at the SALR on saturated ascent; it is raised by condensation and lowered by evaporation and is altered by radiative exchange and is therefore in every way unsuitable as a 'tracer'. Dew-point temperature (T_d) is less unsuitable in that it changes by only about one-sixth of the DALR in unsaturated vertical motion and is unaltered by radiative exchanges so long as the air remains unsaturated. The wet-bulb temperature (T_W) is, unlike T and T_d, unaffected by evaporation or condensation† but it changes in adiabatic ascent or descent and also in radiative exchanges‡. Potential temperature (θ) and mixing

---

† T_W is, by definition, the lowest temperature which a particular air sample can attain by evaporation of water into it. The reverse process occurs on condensation from the air, the latent heat raising its temperature and T_W remaining constant.

‡ T changes but r (and therefore T_d) is unchanged: T_W is thus altered in accordance with Normand's theorem.

ratio (r) are both conserved in unsaturated adiabatic motion but not in saturated ascent: θ, but not r, is altered by radiation. Finally, wet-bulb potential temperature (θ_W), which is derived from T_W, has the added advantage of remaining constant in adiabatic motion but (like T_W) is not conserved in radiative exchanges.

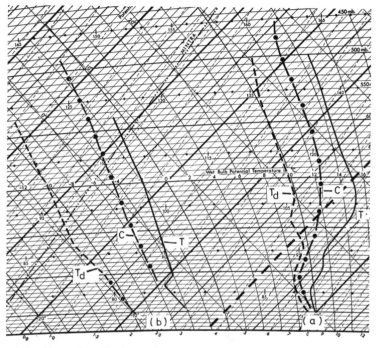

Fig. 5.9. Examples of winter air masses at Crawley, southern England: (*a*) tropical maritime, 12 G.M.T. on 14 January 1968; (*b*) arctic, 12 G.M.T. on 8 February 1969. C is the 'characteristic' curve.

Of these elements (and including vapour pressure and relative humidity as well as other derived humidity parameters we have not defined) r is, in general, most suitable as an air mass indicator when there is no condensation or precipitation, and θ_W is most suitable in other circumstances (air masses differ more in r than in θ_W).

(2) *Classification*

For the British Isles, as for many other parts of the world, extensive data have been collected concerning the properties (in separate seasons) of homogeneous air masses which affect the region. These data may be represented as temperature and humidity profiles which are sometimes combined in a 'condensation curve', or 'characteristic curve', obtained by joining the points which represent the lifting condensation level at

75

each significant point of an ascent. Examples of two contrasting winter air masses at Crawley, England, are illustrated, in fig. 5.9.

(3) *Frontal surfaces*

As a final illustration of the applications of the tephigram, an example of an ascent curve through a frontal zone (see §8.1.4) is shown in fig. 5.10; it is on data of this kind that information about frontal slopes is obtained.

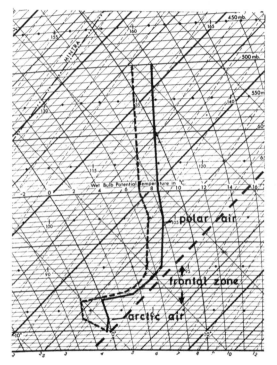

Fig. 5.10. Frontal zone separating arctic and polar air masses: Crawley, southern England, 2 April 1968.

Questions on Chapter 5

5.1. Construct from first principles an adiabatic diagram whose abscissa is a linear scale of temperature (ranging from $-50°C$ to $+30°C$) and whose ordinate is log pressure (increasing downwards from 400 mb to 1000 mb). Draw on the diagram selected dry adiabatics, r_w lines and saturated adiabatics.

5.2. Determine the 1000 mb to 700 mb thicknesses for the arctic and tropical air masses shown in fig. 5.9. What is the vertical thickness of the frontal zone shown in fig. 5.10?

5.3. Calculate the values of 'adiabatic' water content of clouds, at 700 mb and 500 mb, formed by convection from the Earth's surface (1000 mb) under the following four sets of conditions: $T = 20°C$, $T_d = 16°C$ and $12°C$; $T = 18°C$ and $25°C$, $T_d = 16°C$.

5.4. Plot the following environment curves: $p = 1000$ mb, $T = 5°C$, $T_d = 1°C$; $p = 900$ mb, $T = 3°C$, $T_d = -1°C$; $p = 800$ mb, $T = 0°C$, $T_d = -2°C$; $p = 700$ mb, $T = -6°C$, $T_d = -12°C$. Find the values of r, e, T_W, U, θ and θ_W for the air at 900 mb: what are the values of each after (1) ascent by 50 mb and (2) descent by 50 mb? Identify any potential instability in the curve and state the extent of the vertical motion required to release it.

5.5. Three environment curves with the lapse rates respectively dry adiabatic, saturated adiabatic and isothermal pass through the point 800 mb, 2°C. Find graphically the temperature change at 800 mb for a descent of 100 mb in each environment. Suggest a general formula for local rate of change of temperature $(\partial T/\partial t)$ owing to subsidence, on the basis of these cases and other considerations.

5.6. A saturated air stream of lapse rate corresponding to $\theta_W = 15°C$ crosses a range of mountains 1 km high at a mean speed of 40 knots. If it is assumed that the surface air ascends by 1 km and that the ascent decreases in proportion to height to zero at 3 km, calculate the rate of rainfall: assume that all the condensed water vapour is precipitated on the upslope which is 100 km in extent.

CHAPTER 6
winds

THE air clings to the Earth's surface by gravity and, to a large extent, rotates with it. The horizontal air motions which we observe and recognize as winds are those which relate to a fixed position on the spinning Earth. Paradoxically, winds would hardly exist if the Earth did not rotate.

Winds are caused by pressure forces in the horizontal and, for the most part, result from a near-balance between this force and one other, called the Coriolis force, which arises from the Earth's rotation. Other forces are important in certain circumstances, notably when the air's path is very curved or is within the friction layer near the Earth's surface.

The horizontal pressure forces change with height in the atmosphere and winds change correspondingly. The shear vector which represents a particular change of wind velocity is controlled by the horizontal gradient of temperature in the layer concerned and is therefore called the thermal wind. The concept of the thermal wind readily explains why, on average, the westerly component of wind increases with height within the troposphere nearly everywhere.

The temperature distribution in a particular region sometimes favours the appearance of a restricted belt of very strong winds, called a jet stream, usually in the high troposphere. Jet streams are important in themselves, in aviation particularly, and also for their dynamical effects.

6.1. Laws of motion and the Earth's rotation

6.1.1. Newton's first and second laws

According to Newton's Second Law of Motion the action of a single force on a particle is to cause it to accelerate in the direction of the force. When two or more forces act, then the particle accelerates in the direction of their resultant. If there is no resultant force there is no acceleration, i.e. the particle either remains at rest or in uniform motion in a straight line (First Law).

The two main forces acting on the air are the vertical pressure gradient force (p.g.f.) and the opposing force of gravity (see § 2.2.3). When these forces do not exactly balance, the air accelerates vertically for a short time and the initial imbalance is soon redressed (unaccelerated vertical motion usually continues for some time after balance is regained).

The existence of p.g.f.'s in the horizontal plane is apparent from the isobaric patterns on synoptic charts: they are always very much smaller than the vertical p.g.f.'s but have no obvious counteracting force. The question therefore arises as to how such pressure gradients are able to persist, and even intensify, when the 'expected' motion of the air (namely, an acceleration in the direction of the unbalanced p.g.f., i.e. from high to low pressure normal to the isobars) should cancel pressure differences almost as soon as they arise. Inspection of a weather map shows that the air motion is not, in fact, at right angles to, but is (almost) along, the isobars. The persistence of pressure systems is understandable in the light of this motion but a basic question remains: why does the air move in this way, in apparent contradiction to Newton's Second Law? It turns out that there is no contradiction when account is taken of the fact that the surface rotates under the moving air.

6.1.2. *Nature of the Earth's rotation*

All points of the solid Earth rotate in a west–east direction about the polar axis and complete one revolution (2π radians) in a day†. The angular velocity vector, denoted Ω, is illustrated in fig. 6.1 at various points of the Earth: the usual convention for such a vector applies, namely that the fingers of the right hand indicate the sense of rotation when the thumb points in the direction of the vector. The plane of the rotation is at right angles to the vector direction.

Although Ω is the same at all points of the solid Earth, fig. 6.1 shows that its significance changes greatly with latitude. The vector is in the local vertical plane at both poles and therefore represents a rotation in the horizontal plane there. However, since Ω points into the Earth at the south pole and out of it at the north pole the rotations relative to the Earth are opposite in sense, namely clockwise from above the south pole and counter-clockwise viewed from above the north pole. At the equator, on the other hand, Ω lies in the horizontal plane and therefore there is no component of rotation about the local vertical.

For our purposes it is helpful to resolve Ω into its horizontal and vertical components: simple geometry shows that these are $\Omega \cos \phi$ and $\Omega \sin \phi$ respectively, where ϕ is the latitude. Since, by convention, ϕ is reckoned positive when measured counter-clockwise from the line representing the equatorial plane, $\sin \phi$ has opposite signs in the two hemispheres. The horizontal component $\Omega \cos \phi$, on the other hand,

† More accurately, the Earth's angular velocity is $2\pi \times \dfrac{366\frac{1}{4}}{365\frac{1}{4}}$ rad day^{-1} $= 7 \cdot 29 \times 10^{-5}$ rad s^{-1} because the Earth makes one extra revolution (round the sun) in a west–east direction in the course of a year.

has the same sense (towards local north) everywhere. Note also that when $\phi = 0°$ (equator) $\Omega \sin \phi = 0$ and $\Omega \cos \phi = \Omega$, and that when $\phi = \pm 90°$ (poles) $\Omega \sin \phi = \Omega$ and $\Omega \cos \phi = 0$.

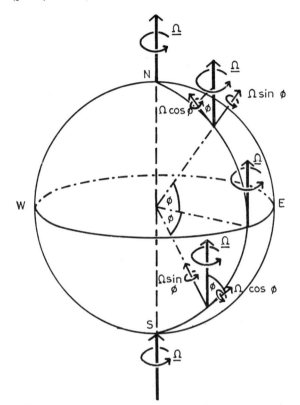

Fig. 6.1. The Earth's angular velocity vector Ω shown at various latitudes, with vertical and horizontal components.

The horizontal and vertical components of Ω are generally of about the same magnitude but their effects on air motion are not at all comparable in importance. $\Omega \cos \phi$ represents the Earth's rotation in the local vertical plane where it has to compete with large pressure gradient and gravity forces and therefore has negligible effects. The $\Omega \sin \phi$ component, on the contrary, has vital effects on the winds.

6.1.3. *Effects of the Earth's rotation: the Coriolis force*

(1) *Some effects and illustrations*

A simple illustration of the effect which is involved when air moves horizontally across the Earth's surface is as follows. Draw a line, at a steady speed and in a fixed direction, on a sheet of paper which is steadily rotated. Such a line is seen to have a curvature opposite to the direction in which the paper is rotated, i.e. the curvature is clockwise if the paper is rotated counter-clockwise, and vice versa. It is easily confirmed also that the degree of curvature of the line depends on the relative speeds of rotation and of drawing: the curvature is small if the paper is turned slowly and the line drawn quickly, and it is large if the opposite conditions apply. It is also easily confirmed that the results obtained are the same for lines drawn outwards from the axis of rotation and inwards from the edge of the paper.

The same basic effect arises in the motion of all bodies across the Earth's surface. When the distance travelled is small and the speed is large the curvature effect is negligible as, for example, in rifle fire at a nearby target. When a short distance of travel at low speed is involved the effect may not, in principle, be negligible but is usually made so by the effect of friction on the moving particle—see, however, problem 6.3. At large distances, allowance for the curvature of the path is required even at fairly high speeds as, for example, in long-range gunnery and in aviation (in the latter case by flying on a fixed compass bearing).

While all motion over the Earth is affected by the Earth's rotation it very often turns out, in everyday life, that the effect is quite negligible. Thus, the turn-table mentioned in problem 6.1 rotates once in 10 seconds relative to the Earth which itself rotates once in 24 hours: if we neglect the latter motion in this problem we omit a correction factor of only 1 in 8640. The tacit assumption that we may neglect the Earth's rotation is, in fact, also involved in the laboratory verification of Newton's Second Law. In meteorology, however, it is only near the equator, and in very local wind systems elsewhere, that the Earth's rotation fails to play a very important part.

(2) Deviation of the air: the Coriolis force

We may think of air, which is moving horizontally in latitude ϕ, as if it were traversing a turn-table which is coincident with the tangent plane and is rotating about the local vertical with angular velocity $\Omega \sin \phi$.

This is illustrated in fig. 6.2, in which the northern hemisphere condition (counter-clockwise rotation of the Earth) is assumed. The air is initially moving from O towards a point P on the Earth. P is rotating counter-clockwise and the air therefore turns clockwise with respect to it, i.e. to the right. If we assume steady conditions in which the air moves with constant velocity V, then after time t it has moved distance $d = Vt$; meanwhile P has rotated from P_1 to P_2 where the angle $P_1 O P_2 = \Omega \sin \phi . t$. The air is now at a distance s from P given by

$$s = P_1 P_2 = Vt \times \Omega \sin \phi . t$$

$$= V\Omega \sin \phi . t^2.$$

81

The curved path of the air relative to the ground is shown in fig. 6.2†.

The transverse displacement of the target relative to the air (i.e. its displacement in a direction sideways from the velocity vector) increases steadily from zero at O to s at P_2. The air may be considered

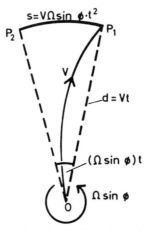

Fig. 6.2. Horizontal air motion, in northern latitude ϕ, relative to the Earth.

to be subject to a uniform transverse acceleration, a, or to the corresponding transverse force, which causes its path to be curved. We find what this acceleration is, in terms of the various quantities concerned, by comparing the result

$$s = V\Omega \sin \phi . t^2 \tag{6.1}$$

with a standard equation in dynamics

$$s = ut + \tfrac{1}{2} at^2$$

which, since the initial transverse velocity u is zero, reduces to

$$s = \tfrac{1}{2} at^2. \tag{6.2}$$

Comparison of equations (6.1) and (6.2) gives $a = 2\Omega \sin \phi . V$. This is also the magnitude of the transverse force acting per unit mass of air (since force = mass × acceleration) which is usually called the Coriolis force‡ after its 19th century discoverer: its magnitude, $2\Omega \sin \phi . V$ is written fV for short ($f = 2\Omega \sin \phi$ is the 'Coriolis parameter'). The force increases in direct proportion to $\sin \phi$ and to the wind velocity.

† The curvature of the path is produced, not in the large finite step portrayed, but in a series of infinitesimal steps in which the difference between the arc and radial distances (OP_1 of fig. 6.2) is negligible.

‡ It is also often called the geostrophic (literally, earth-turning) force, sometimes the deviating force.

It always acts at right angles to the wind and therefore affects its direction but not its speed. Although it is an apparent force in the sense that it arises only because the air moves over a rotating surface it affects in a fundamental way all the meteorological elements.

6.2. *Inertial flow and geostrophic winds*

6.2.1. *Nature of inertial flow*

All bodies, including air, which move horizontally across the Earth's surface are subject to the Coriolis force (C.f.) which, unless compensated, causes them to curve continuously to the right in the northern hemisphere and to the left in the southern hemisphere. The path of a body which moves across the Earth under the action of no external forces is therefore circular and the radius of the circle depends on the velocity and latitude†—see problem 6.3. In meteorological terms this is the ' circle of inertia '. Pure inertial motion rarely arises because there is nearly always at least a small horizontal p.g.f.; further, close to the equator where the p.g.f. is most liable to be absent or very weak, the C.f. force itself is absent (since $f = 2\Omega \sin \phi = 0$) and thus the circle of inertia condition of flow does not apply there.

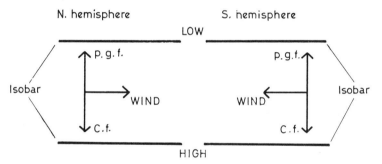

Fig. 6.3. Geostrophic equilibrium and wind in northern and southern hemispheres.

6.2.2. *Nature of geostrophic flow*

Since a horizontal p.g.f. is needed to move the air and since the air, once set in motion, is subject to the C.f., the simplest atmospheric motion conceivable is one in which these two forces alone are involved in such a way that they balance exactly, because there is then no acceleration to complicate the motion. The only possible arrangements of wind and forces (different in the two hemispheres) are shown in fig. 6.3 and are explained as follows. The p.g.f. acts at right angles to the

† The Coriolis force per unit mass (fV) produces a centripetal acceleration (V^2/r) so that $r = V/f$: see also equation (6.7) and the paragraph following it. The path of the body is not quite circular, especially in low latitudes, because of the variation of f with latitude.

isobars, from high to low pressure; the C.f. is equal and opposite to the p.g.f.; since the C.f. is also at right angles to and to the right (left) of the wind in the northern (southern) hemisphere the air must move along the isobars with low pressure to the left in the northern hemisphere and to the right in the southern hemisphere. These directions for the geostrophic wind agree with observations of the actual wind, as expressed in Buys Ballot's law of the mid-19th century.

Geostrophic balance is a state which nearly holds (almost) everywhere in the atmosphere but does not exactly hold anywhere. This remark serves to emphasize both the usefulness of the geostrophic wind concept and its restrictiveness. Its practical use in the estimation of wind velocity will be demonstrated; it is also a useful simplifying assumption in theoretical work. However, it strictly implies no acceleration or vertical motion or pressure change and its usefulness is therefore limited in the study of large-scale pressure systems and rain-producing mechanisms.

6.2.3. *Geostrophic wind equation*

The geostrophic balance of forces acting per unit mass is expressed in the following equation, in which $-1/\rho(\partial p/\partial n)$ is the p.g.f. (see § 2.2.5) and fV_G is the C.f.:

$$-\frac{1}{\rho}\frac{\partial p}{\partial n}+fV_G=0,$$

$$\text{i.e.} \quad V_G=\frac{1}{f\rho}\frac{\partial p}{\partial n}. \tag{6.3}$$

This simple equation, or the equivalent form given in equation (6.4), is very important because it allows a good estimate to be made of wind speed when no direct observations of winds are available. The method of using equation (6.3) to deduce wind speed from isobar spacing may be illustrated as follows. Figure 6.4 shows a supposed distribution of mean-sea-level isobars†, drawn at 4 mb intervals and at different distances apart. The values of $\partial p/\partial n$ at A, B and C are (as always in calculations in meteorology) approximated by finite differences, $\Delta p/\Delta n$; for example:

$$\left(\frac{\Delta p}{\Delta n}\right)_A=\frac{4}{10^2}\text{ mb km}^{-1}=\frac{4\times 10^2}{10^2\times 10^3}\text{ S.I. units.}$$

If the latitude is assumed to be 60°, say, then

$$f=2\Omega\sin\phi=1{\cdot}37\times 10^{-4}\text{ s}^{-1};$$

if, further, we assume an air density of $1{\cdot}2$ kg m^{-3}, then

$$(V_G)_A=\frac{4\times 10^{-3}}{1{\cdot}37\times 10^{-4}\times 1{\cdot}2}=24\text{ m s}^{-1}\ (\approx 48\text{ knots}).$$

† Lines of constant pressure drawn to fit a network of simultaneous pressure observations each adjusted to mean sea level.

Similarly $(V_G)_B = 12$ m s^{-1} (the speed is reduced in the same proportion as the isobar spacing increases); also $(V_G)_C = 16$ m s^{-1} (in this case the most appropriate value of $\Delta p/\Delta n$ is (8/300) mb km^{-1}).

A value of V_G so obtained from mean-sea-level isobars cannot be directly applied to the surface level because of the frictional retarding force exerted on air motion near the ground. It gives, however, a good approximation to the wind at the level just above the friction layer if, as is usually assumed, the isobar spacing alters little between the ground and this level.

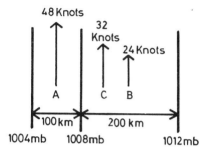

Fig. 6.4.   Pressure gradients and calculated geostrophic winds.

The measurement of V_G at various points in a system of isobars is made easier by the use of a ' geostrophic wind scale '. This is usually of Perspex and is graduated in units of speed (knots), corresponding to selected spacings of isobars on a specified scale of chart. Adjustment to the indicated speed may be required for air density or for scale of chart; usually, provision is made in the scale for measuring the speed at different latitudes.

At any given latitude, the wind speed is greater in direct proportion to the closeness of adjacent isobars. The effect of latitude should, however, be borne in mind in visual inspection of isobars over a large area: for example, with a given pressure gradient the wind speed in high latitudes is only about half that in latitude 30° (the effect of different densities is smaller but is in the same sense). The physical explanation is in line with that given earlier and is as follows. The curvature of the path of air, moving at a given speed, is proportional to $\sin \phi$ and so is twice as great near the pole as at 30°. The p.g.f. required to balance the curvature effect and so ' straighten ' the motion is therefore also twice as great near the pole as it is at 30°. We examine later how curvature of the motion is produced by a lack of balance between the p.g.f. and C.f.

The geostrophic wind equation has an alternative form to equation (6.3) when the p.g.f. is expressed in terms of the slope of the isobaric surface, i.e. p.g.f. $= -g\partial Z/\partial n$ (see equation (2.14)). The geostrophic wind equation is then

$$V_G = \frac{g}{f} \frac{\partial Z}{\partial n}.$$ (6.4)

Constant level (mean-sea-level) analysis of pressure has been used for surface observations since the systematic study of meteorology began in about the mid-19th century. On the other hand, constant pressure analysis was adopted in the regular study of conditions in the upper atmosphere over an extensive area almost from the time (early 1940's) that radiosonde observations first made this possible. Despite the obvious awkwardness that the 500 mb level, say, does not refer to a definite height above sea level (but only to about 5·5 km, the exact

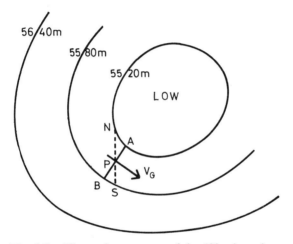

Fig. 6.5. Illustrative contours of the 500 mb surface.

height depending on the sea-level pressure and on the average air temperature in the column—see altimeter equation (2.10) or (2.11)), the use of constant pressure (isobaric) analysis does not amount simply to perverseness on the part of the meteorologist. There are advantages to him in that many of the dynamical equations which he uses are simpler and easier to interpret in their 'constant pressure' form. An immediate practical gain is that a single geostrophic wind scale based on equation (6.4) can be used on a chart at any pressure level, the variation of g with height in the atmosphere being insignificant; whereas allowances are required—different at each level—to account for the variation of ρ with height if a scale based on equation (6.3) is used.

Figure 6.5 is a plan view of typical 500 mb contours over a small area; fig. 6.6 is a cross-section along the line of maximum slope (AB) at point P of fig. 6.5†. The 500 mb surface intersects the contours 5520 m and 5580 m at points C and D, and points A and B are the projections of C and D on to the mean-sea-level surface. In fig. 6.6, the contours are normal to the diagram while V_G (blowing so as to keep the lowest pressure, at the same level, to the left) is parallel to the contours with low values to the left. The same direction of V_G is shown in fig. 6.5. The direction of V_G is reversed in the southern hemisphere.

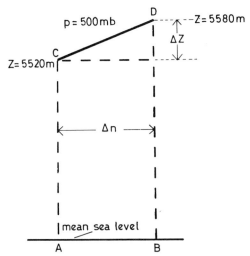

Fig. 6.6.   Cross-section along the line of greatest slope at point P on fig. 6.5 (looking downwind).

The very gentle slopes of pressure surfaces are much exaggerated in fig. 6.6.   Since, in round figures, for a slope of 1 in 100 000 in middle latitudes,

$$V_G = \frac{g}{f} \frac{\Delta Z}{\Delta n}$$

$$= \frac{10}{10^{-4}} \times 10^{-5} = 1 \text{ m s}^{-1},$$

and since V_G is proportional to slope, then the normal range of variation of slope is about 10^{-3} to 10^{-5} (wind speeds up to about 100 m s^{-1} or 200 knots).

6.2.4. *Wind and pressure near the equator*

As the equator is approached the geostrophic balance between pressure and wind, which in general holds well in higher latitudes,

† The shortest distance between adjacent contours: any other line corresponds to a component of V_G, e.g. the north–south line NS relates to the west–east component.

breaks down because the deflecting force caused by the Earth's rotation about the vertical tends towards zero there. Tropical cyclones move away from the equator; the horizontal pressure variations close to the equator are small; and such pressure fields as exist do not control the air motion in any simple way. The prevailing conditions are of light, variable winds ('doldrums'). These facts show how vital the Coriolis force is in the development of the large pressure and wind fields which are observed over most of the earth.

6.3. *Gradient winds*

If the wind is somewhat stronger than is appropriate to geostrophic balance then the C.f., being proportional to the wind speed, exceeds the p.g.f. Since the C.f. acts to the right of the motion in the northern hemisphere the air curves to the right; and since high pressure is to the right of the wind, the flow is concave to high pressure and is said to have anticyclonic curvature (a circulation completely enclosing high pressure being called an anticyclone). As we saw earlier, the C.f. produces a rotation opposite to that of the Earth which must therefore have a rotation in the opposite, cyclonic, sense: the same argument holds for the southern hemisphere, with 'left' replacing 'right'. The anti-cyclonic sense of rotation is clockwise in the northern hemisphere, counter-clockwise in the southern hemisphere, each opposite to that of the Earth (see § 6.1.2).

The opposite situation in which the wind is not strong enough to sustain geostrophic balance need not be argued in full. What happens, clearly, is that the Earth imparts some of its own (cyclonic) rotation to the air and so the curvature of the flow is reversed compared to the previous case. Again, it is clear that a curved circulation of this kind will be able to persist only if the isobars are (almost) parallel to the wind, since any other situation implies the action of distorting forces on the flow. Balanced conditions, in which the net force acting on the air is normal to the wind and to the isobars and produces an inward (centripetal) acceleration in the direction of the stronger force, are assumed in fig. 6.7. Balanced winds of this kind, blowing along curved isobars, are called gradient winds (V_{gr}).

The balanced conditions of fig. 6.7 are represented by two quadratic equations in which the net force acting on the air is equated to the inward acceleration, opposite to the p.g.f. and therefore positive in the anticyclonic case, and negative in the cyclonic case.

Anticyclonic: $$-\frac{1}{\rho}\frac{\partial p}{\partial n}+fV_{gr}=+\frac{V_{gr}^2}{r}; \qquad (6.5)$$

Cyclonic: $$-\frac{1}{\rho}\frac{\partial p}{\partial n}+fV_{gr}=-\frac{V_{gr}^2}{r}. \qquad (6.6)$$

Since $fV_G = \dfrac{1}{\rho}\dfrac{\partial p}{\partial n}$ from equation (6.3),

$$V_{gr} = V_G \pm \frac{V_{gr}^2}{fr}. \tag{6.7}$$

The second term on the right-hand side is called the cyclostrophic term; its importance increases as r decreases, and for $r = \infty$ the expected result that $V_{gr} = V_G$ is obtained. The term is positive for anticyclonic curvature ($V_{gr} > V_G$) and negative for cyclonic curvature ($V_{gr} < V_G$),

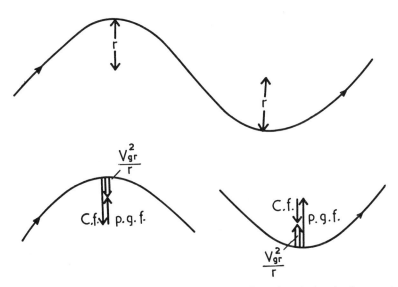

Fig. 6.7.   Balance of forces in curved flow (northern hemisphere); flow anti-cyclonic, then cyclonic.

in accordance with our starting premises. We may note, too, in passing that, if in equation (6.7) $V_G = 0$ (this implies no p.g.f.) and the positive sign is taken, then anticyclonic motion (curvature to the right in the northern hemisphere) with a radius of curvature of $r = V_{gr}/f$ is implied: this is the inertial motion discussed in § 6.2.1.

Equations (6.5) and (6.6) may be solved separately, the necessary condition for non-inertial motion (i.e. $V_{gr} = 0$ when $(1/\rho)(\partial p/\partial n) = 0$) being applied in order to select the appropriate solution of the two which are obtained for each equation. The student may confirm that the correct solutions under these conditions are as follow:

Anticyclonic: $\qquad V_{gr} = \dfrac{fr}{2} - \sqrt{\left(\dfrac{f^2 r^2}{4} - \dfrac{r}{\rho}\dfrac{\partial p}{\partial n}\right)};$ $\qquad$ (6.8)

89

Cyclonic:
$$V_{gr} = -\frac{fr}{2} + \sqrt{\left(\frac{f^2 r^2}{4} + \frac{r}{\rho}\frac{\partial p}{\partial n}\right)}. \tag{6.9}$$

The most interesting points about these solutions are that there is a maximum possible speed, namely $fr/2$, in anticyclonic rotation but there is no maximum speed in cyclonic rotation (though, in practice, friction will eventually impose a maximum in this case). The reason for this difference is clear. The Earth itself has cyclonic rotation: it imparts this to the atmosphere in general and aids the development of local cyclonic circulations. Parts of the atmosphere, however, rotate at a speed less than that of the Earth: they appear to us as anticyclones but they are limited by the Earth to a speed which keeps the air's rotation in space just cyclonic. (The maximum anticyclonic rotation, $fr/2$, at a radius of curvature r, represents an angular velocity of $f/2$, i.e. $\Omega \sin \phi$; this is the same as the Earth's angular velocity but in the opposite sense.)

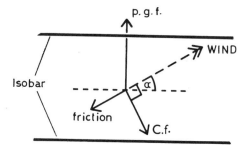

Fig. 6.8. Balance of forces within the friction layer (northern hemisphere): C.f. is normal to the wind which is backed by angle α from the isobars.

6.4. *Winds in the friction layer*

In the lowest 0·5 km or so of the atmosphere the Earth's surface exerts a frictional drag on the air and reduces the wind speed, the greatest reduction being near the ground. More is said about this process in Chapter 9. Here we show in fig. 6.8 how the geostrophic balance is altered by the additional frictional force, which is assumed to act against the wind. The reduction of wind speed causes the C.f. to fall below the p.g.f. so that the air acquires a component across the isobars to low pressure. Equilibrium is established with the frictional force equal and opposite to the resultant of the other two forces.

The surface wind velocity is measured at a standard height of 10 m and is, on average, backed from the isobars (i.e. changed in a counter-clockwise sense)† by about 30° over the land and about 10° over the sea. Its speed is about one-third of V_G over the land and

† The change is in the opposite sense in the southern hemisphere.

about two-thirds of V_G over the sea. When surface air over the land interchanges freely with the faster-moving air aloft, the ratio of the speeds is larger, and the angle of backing is less, than these average values. The opposite effects occur when the air is stable and there is little or no vertical mixing. Diurnal stability fluctuations are often pronounced over the land and account for the noticeable tendency for the surface wind to freshen to a maximum (and become more gusty) in the afternoon and decrease during the night. When the surface layers are stable the higher parts of the friction layer would be expected to move more freely than when vigorous mixing carries slower-moving air up from the ground. Wind measurements made on high towers have confirmed this by showing that the average diurnal changes of wind speed and direction there are opposite to those near the ground.

6.5. *Thermal winds*

6.5.1. *Vertical shear vector*

Changes of wind velocity within the free atmosphere are of quite a different nature from those in the friction layer. The vertical wind-shear vector $\mathbf{V}_S$ is defined by the equation

$$\mathbf{V}_S = \mathbf{V}_2 - \mathbf{V}_1$$

in which the wind vectors $\mathbf{V}_1$ and $\mathbf{V}_2$ apply to the lower and upper levels, respectively. When $\mathbf{V}_1$ and $\mathbf{V}_2$ are drawn (in magnitude and direction) from a common origin, as in fig. 6.9, $\mathbf{V}_S$ is obtained by joining the end point of the lower vector to that of the upper vector. In the example shown, a south-easterly at the lower level changes to a south-westerly at the upper level by the addition of a predominantly westerly shear vector.

6.5.2. *Temperature control of the shear vector*

Since the upper winds are (nearly) geostrophic and usually change with height, the distribution of pressure must, in general, be different

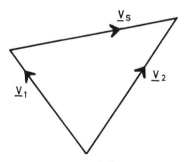

Fig. 6.9. Vertical shear vector.

from one atmospheric level to another. We shall therefore seek to explain the wind changes in terms of the corresponding horizontal p.g.f. changes (or the equivalent changes of slope of isobaric surfaces). It simplifies matters if we assume, in the first instance, that there is no horizontal p.g.f. at the Earth's surface which is taken to be mean sea level; the actual wind at any height is then simply the shear vector between this height and mean sea level. (More generally, since $V_2 = V_1 + V_S$, the mean-sea-level wind has to be added to the shear vector to obtain the wind at the higher level: in this addition the effect of friction on the mean-sea-level wind is eliminated by assuming the latter to be given either by the wind at the top of the friction layer or by the mean-sea-level pressure gradient.)

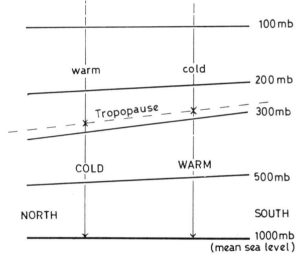

Fig. 6.10. Schematic changes of slope of isobaric surfaces in middle latitudes resulting from the average distribution of temperature in the northern hemisphere. The mean isotherms are assumed to be normal to the paper, i.e. parallel to the latitude lines.

Change of pressure with height is solely determined by temperature and decreases more through a cold layer than through a warm layer of the same thickness; this also implies that the separation of two isobaric surfaces decreases with temperature. Since, on average, cold air lies on the poleward side at all levels in the troposphere (see fig. 2.2) there is a horizontal p.g.f., directed from low to high latitudes and increasing with height, upwards to the tropopause; thus the westerly component of wind increases with height (low pressure being to the left of the wind in the northern hemisphere). Above the tropopause the temperature gradient is generally in the opposite sense† so that the westerly component initially decreases with height and may change to

† Not near the winter pole—see fig. 2.2.

easterly. Thus in fig. 6.10 the 500 mb surface is represented as tilting down towards the pole and the 300 mb surface more so, while the reverse trend is shown above the tropopause. Since the mean isotherms are assumed in fig. 6.10 to be parallel to the latitude lines, i.e. normal to the paper, then the general rule applies that the shear vector in geostrophic flow is parallel to the mean isotherms of the layer concerned and has low temperature on the left in the northern hemisphere. It is also clear that the larger the horizontal gradient of temperature the bigger is the difference of slope of adjacent isobaric surfaces and the stronger is the shear through a given layer.

Because of the reversed sense of the relationship between pressure and wind direction in the southern hemisphere the geostrophic shear vector has low temperature on the right there; however, this is easily seen to imply the same directions of shear (westerly in the troposphere, easterly above) as in the northern hemisphere.

6.5.3. *Thermal wind equation and thickness charts*

The shear vectors which are deduced from geostrophic considerations are called thermal winds ($\mathbf{V}_T$) because of their control by the temperature distribution. They are usually very nearly the same as the actual shear vectors ($\mathbf{V}_S$) and are often identified with them, just as geostrophic winds are identified with actual winds in the free atmosphere.

The 'thermal wind equation' is obtained in its constant pressure form by applying equation (6.4) to both the upper and lower isobaric surfaces (2 and 1) which define a particular layer, and then subtracting. Thus

$$\mathbf{V}_{G2}=\frac{g}{f}\frac{\partial Z_2}{\partial n} \quad \text{and} \quad \mathbf{V}_{G1}=\frac{g}{f}\frac{\partial Z_1}{\partial n}$$

$$\therefore \quad \mathbf{V}_{G2}-\mathbf{V}_{G1}\equiv\mathbf{V}_T=\frac{g}{f}\frac{\partial(Z_2-Z_1)}{\partial n} \tag{6.11}$$

The thickness of the isobaric layer concerned, (Z_2-Z_1), is directly proportional to the mean temperature of the layer (see equation (2.11)). Equation (6.11) therefore signifies that the thermal wind speed is directly proportional to the gradient of layer thickness or, equally, of layer mean temperature. Similarly, the wind direction is parallel to the mean isotherms or thickness contours.

The values of thickness of a given isobaric layer, say 1000 to 500 mb, are easily calculated for upper-air stations and a set of contours ('thickness lines') may be drawn for a geographical area on the basis of these. The chart concerned is a 'thickness chart' and it represents in a very convenient way the geographical distribution of mean temperature in the layer concerned. This is a property of the atmosphere which is especially important not only in relation to wind shear but also to the development and movement of pressure systems (see Chapter 8). The thermal wind speed is measured by the usual geostrophic wind scale; its direction, parallel to the contours, has low thickness to the left in the northern hemisphere.

It is obvious that the three charts 1000 mb, 500 mb and 1000 to 500 mb thickness are interdependent in the sense that when the contours of any two are fixed those of the remaining chart are known: for example, the 1000 mb contours are obtained by subtracting the 1000 to 500 mb thickness contours from those for 500 mb (most conveniently by overlaying one set of contours with the other and joining appropriate intersections). This interdependence acts as a useful cross-check in the prediction of contours for, while all three sets of contours may to a certain extent be predicted independently, the patterns must at all times be mutually compatible.

Although the professional meteorologist usually works in terms of layers bounded by isobaric surfaces it is sometimes convenient to use an approximate expression for the thermal wind speed given in terms of the horizontal temperature gradient and the layer depth in height units. The required equation is

$$\mathbf{V}_T = \frac{g}{f\bar{T}} \frac{\Delta \bar{T}}{\Delta n} (Z_2 - Z_1). \qquad (6.12)$$

From equation (6.12) the numerical value of $\mathbf{V}_T$ is found to be 3·6 m s^{-1} (7·2 knots) for a layer 1 km deep and a horizontal temperature gradient of 1 K/100 km, in middle latitudes ($f = 10^{-4}$ s^{-1} and $\bar{T} = 270$ K, say). For a given value of $\Delta \bar{T}/\Delta n$, $\mathbf{V}_T$ is thus proportional to the layer depth—hence the usual increase of wind with height in the troposphere; and for a given depth of layer $\mathbf{V}_T$ is proportional to $\Delta \bar{T}/\Delta n$—hence the strong upper winds often found near fronts.

6.5.4. *Hodographs and temperature advection*

If wind velocities at selected levels, measured at a particular place, are plotted on a polar diagram and the end points of the vectors for adjacent levels are joined, then the resulting curve (in practice a series of straight lines) is called a (wind) hodograph. Only the end point of each vector need be plotted, as illustrated in fig. 6.11 *a* for levels 1 to 5 corresponding, say, to 1 to 5 km: the wind vectors concerned are all from the south-west quadrant and their end points, correspondingly, are in the north-east quadrant†.

For illustration, the wind vectors for the levels 1 and 2 km are drawn in fig. 6.11 *b*. The magnitude of $\mathbf{V}_S$ in the layer is found by measurement to be 11 knots; the same value may be calculated by subtracting the south–north and west–east components of $\mathbf{V}_1$ from those of $\mathbf{V}_2$ and

† In an alternative convention the wind vectors are drawn so as to end at the origin, $\mathbf{V}_S$ then being directed from the starting point of the higher vector to that of the lower vector.

94

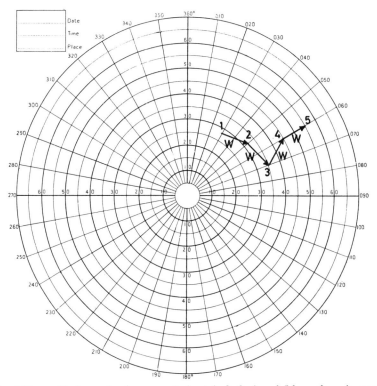

Fig. 6.11 *a*. Hodograph of upper winds at 1, 2, 3, 4 and 5 km: there is warm air advection from 1 to 3 km, cold air advection from 3 to 4 km and no temperature advection from 4 to 5 km.

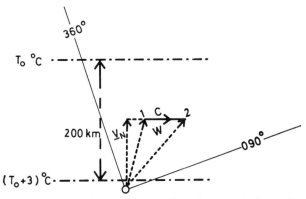

Fig. 6.11 *b*. Calculation of warming rate by advection in layer 1 to 2 km of fig. 6.11 *a*. The shear vector corresponds to a temperature gradient of 1·5 K/100 km.

95

recombining. If we identify $\mathbf{V}_S$ with $\mathbf{V}_T$ then the colder and warmer air are distributed as indicated; and from equation (6.12) a value for $\Delta T/\Delta n$ of $1 \cdot 5$ K/100 km is obtained on the basis of the previous assumptions concerning f and T. This gradient of mean temperature in the layer is indicated in fig. 6.11 b by the positioning of isotherms. Since the winds at the bottom and top of the layer (and, by inference, throughout it) are from the south-west, warmer air is replacing colder air at the station in this layer. If the wind component normal to $\mathbf{V}_T$ (i.e. parallel to the temperature gradient) from the 'warm' side were 100 km hr^{-1} then obviously T in the layer would be rising at the rate $1 \cdot 5$ K hr^{-1}. The actual rate is given by:

$$\frac{\Delta T}{\Delta t} = \frac{\Delta T}{\Delta n} \times \mathbf{V}_N \qquad (6.13)$$

where $\mathbf{V}_N$ is the component of $\mathbf{V}_1$ or $\mathbf{V}_2$ normal to $\mathbf{V}_T$ (these components are equal). In this example, $\mathbf{V}_N$ is measured as 27 knots and thus $\Delta T/\Delta t$ is $0 \cdot 4$ K hr^{-1}.

This calculation serves to illustrate how the instantaneous rate of 'temperature advection', i.e. horizontal transport of warmer or colder air at a particular place, may be inferred from a hodograph. In the northern hemisphere winds that veer with increasing height signify warm air advection while backing winds are easily seen to imply cold air advection (as between 2 and 4 km in fig. 6.11 a): the association of the letters v and w, and of b and c in this rule is a useful aid to memory.

The processes of large-scale vertical motion, convection and radiation also operate, in general, to change the local temperature distribution. In addition, advection is often important and is of special interest if winds veer at lower levels and back at higher levels, or vice versa, because a change of static stability of the air by advection is then implied (decreasing stability, by warming at low levels and cooling at high levels, in the former circumstance). This is called differential advection.

6.5.5. Jet streams

When deep air masses of contrasting temperature are adjacent along a frontal surface, conditions are suitable for very strong thermal winds. The temperature change across a well-marked front is of the order of 10°C over a horizontal distance of 100 km† and extends from the surface to the tropopause, i.e. through about 13 km in middle latitudes. The corresponding magnitude of $\mathbf{V}_T$, derived from equation (6.12), is over 900 knots and this would apply if the frontal surface were vertical. However, the effective value of the temperature gradient is only a fraction of the above value because the air mass boundary has a gradual slope (see Chapter 8) which has the effect of easing the gradient

† This intensity is rare in the British Isles but is surpassed in some other parts of the world.

applicable to a deep layer. Even so, very strong and rather concentrated thermal winds occur near fronts and these, with the vector addition of the low-level component, give rise to narrow belts of winds, in the high troposphere, which often exceed 100 knots; speeds of over 300 knots have in fact been measured, but not over the British Isles. The wind direction is not far removed from the line of the associated front and is usually between south-west and north-west.

A narrow belt of strong wind of this kind, just below the tropopause, usually a few thousand kilometres in length and with much lighter winds above and below and on either side, is called a jet stream—in this particular case a frontal jet. The importance of such jet streams is enhanced by the strong and localized accelerations and decelerations of flow often associated with them—see §8.7.2(3).

Very strong westerly winds not associated with fronts occur near the tropopause in the sub-tropics (sub-tropical jet) where the horizontal temperature gradient in the troposphere is most marked, on average (see fig. 2.2); also within the stratosphere near the winter poles (' polar night jet ') where the temperature gradient is not reversed compared to the troposphere, as it is in lower latitudes.

Questions on Chapter 6

6.1. A ball is aimed by a person on a turn-table at a target on the turn-table and 5 m away. If the rotation is clockwise at one revolution per 10 seconds and the horizontal speed of the ball is 5 m s^{-1} how should the ball be aimed in order to hit the target?

6.2. A golf ball is struck perfectly, in still air, towards the ' flag ' 100 m away. The ball hits the ground after 6 sec, having covered a horizontal distance of 100 m. By how much does it miss the target? How does this compare with a rifle bullet which travels 100 m in 0·1 s?

6.3. A man walks across a featureless landscape, under a featureless sky and in calm conditions; he walks on an intended straight course but has no compass. After 1 hour's walking at 6 km hr^{-1} in latitude 30° how far has he deviated from his intended course? Show by geometry that a person who walks at speed V completes a circle of radius $6V/\pi \sin \phi$ in a time which is independent of V.

6.4. The separation of isobars in latitude 60° for a given wind speed on a chart of scale 1 : 5 × 10^6 is 1 cm. What is the corresponding separation in latitude 45° for a wind speed twice as great on a chart of scale 1 : 2 × 10^6? If, in the former case, the isobars are 4 mb apart what is the geostrophic wind speed?

6.5. If, in fig. 6.8, the assumption is made that the frictional force is proportional to wind speed V ($= kV$ where k is a constant), show that $V = f V_G/\sqrt{(k^2 + f^2)}$ and that $\alpha = \tan^{-1} k/f$.

CHAPTER 7

instruments and observations

INSTRUMENTAL records of the weather elements cover little more than 200 years anywhere, while for many parts of the world the period of observation is a good deal less than 100 years. The distribution of observations has been heavily biased towards well-populated land regions. Commercial shipping has provided most of the observations over the oceans and these have been supplemented, since about 1950, by a small number of Weather Ships. While pioneer instrumental observations of the upper atmosphere date back to the later years of the 19th century, routine observations of temperature, humidity and wind from a network of stations covering most of the globe have been made for some 25–30 years only.

Three distinct, though overlapping, needs are met by observations of the weather elements. First, weather plays so vital a part in man's activities that an adequate knowledge of climatology is essential, not only for land regions but also for sea and air routes. Second, the various requirements imposed by the demand for weather forecasts have to be met. Third, the need to understand better the working of the atmosphere as a whole means that observations have to be extended to all regions; these observations employ certain techniques, such as constant-level balloons, and include measurements, such as those of ozone, which are not required to meet either of the first two needs.

So far as observations for climatology and weather forecasting are concerned, a distinction is made between ' climatological stations ', on the one hand, and ' synoptic stations ' (perhaps ten times fewer in number), on the other. At climatological stations, observations are usually made only once per day (though continuous recordings of some elements may be made) and the information gained is presented in the form of monthly means or totals, extreme values, distributions etc. At synoptic stations, routine observations are made with a frequency which depends on the importance of the station concerned—usually hourly but sometimes 3-hourly or 6-hourly; this information has a content designed to further synoptic analysis and weather forecasting and is rapidly channelled to recipients.

Satellites, besides supplementing surface reports of clouds in a most striking way (see Plate 13 and fig. 11.2), provide meteorological measurements of various kinds, and hold promise of revolutionizing certain of them. Other applications of modern techniques in weather forecasting include radar detection of rainstorms, radio location of lightning flashes and the automation of weather stations.

The foregoing relates to professional meteorology. To the person concerned with understanding the factors that determine local weather, it can be most instructive to make and interpret small ' field ' experiments carried out with simple apparatus: examples are quoted in §7.4.

7.1. *Routine surface observations*

In general, methods of measurement of the surface elements which yield highly accurate spot values are not only unnecessary in routine observations, but indeed are undesirable because most of the elements concerned are subject to rapid minor fluctuations which constitute unwanted ' noise ' in data required for climatology and weather forecasting†. More important than refinement of instruments is the problem of their appropriate exposure. The elements change most rapidly with height close to the earth's surface and it is vital that standardized exposures be adopted so that the observations may be comparable as between one station and another. Care is needed also to ensure that instrument exposure does not become affected by new buildings or by the growth of trees, for example—such factors being liable to cause spurious long-term changes in observed elements.

7.1.1. *Pressure*

(1) Mercury barometer

The precision aimed at in pressure measurement is rather high, namely 0·1 mb or 1 part in 10 000. The standard instrument is the mercury barometer which is based on the pressure equation, $p = g\rho h$, discussed in § 2.2. Those used in meteorology are calibrated in millibars, standard values of ρ and g being assumed in the calibration: the values are $\rho = 13595 \cdot 1$ kg m^{-3} and $g = 9 \cdot 80665$ m s^{-2}‡. Corrections have to be applied for departures of the actual temperature of the instrument from the standard temperature assumed in the calibration (i.e. the temperature at which ρ has its standard value and h represents true height units on the instrument scale) and for the difference between the local value and the standard value of g§.

† Maximum gust speed is an exception: an anemometer should be capable of measuring this accurately.

‡ These values are substituted in the pressure equation to convert from older height units (mm or inches) to mb, or vice versa: 1 mm Hg (standard) = 133·3 N m^{-2} = 1·333 mb.

§ This is a latitude correction, variations of g with height above mean sea level being negligible. The Earth is ' flattened ' at the poles by its rotation, and g has its maximum value there and its minimum value at the equator. The latitude (ϕ) correction is $-0 \cdot 00259 p$ cos 2ϕ mb, ϕ being counted positive in both hemispheres, i.e. it is about $-2 \cdot 6$ mb at the equator and $+2 \cdot 6$ mb at either pole.

When the mercury rises in the barometer tube it falls in the cistern and it is the difference in these levels which corresponds to h. In a 'moveable cistern' barometer such as the Fortin, the mercury level in the cistern is adjusted, before each reading, to the fiducial point, this being the zero of the height scale. In the fixed cistern Kew Barometer, which is used in the British Meteorological Office, preliminary adjustment is unnecessary because the true height scale is compressed by a factor which depends on the cross-sectional areas of the tube and cistern.

(2) Reduction to mean sea level

For comparability between different stations, observed station-level pressures are reduced to a standard (mean-sea) level by adding the pressure of the missing air column: the altimeter equation (equation (2.10)) is used, p being the observed pressure, p_0 the required sea-level pressure, and z the height of the station. T, the mean temperature of the missing column, is assumed to be given by the screen temperature† (*not* the barometer temperature) at the time of observation: this is an approximation which is liable to introduce significant errors at higher-level stations.

(3) Aneroid barometer and barograph

Aneroid barometers, though not absolute instruments in that they require initial setting to the correct pressure, are portable and, in their modern form, trustworthy instruments. A metal chamber (or series of chambers) is almost evacuated and the ends kept apart by a metal spring. An increase of pressure causes the ends to approach each other (increasing the tension of the spring) and, one end being fixed, the movement of the other is shown by the movement of a pointer on a scale or by some other means. If a long pen arm is used to record the pressure variations on a chart the instrument is a barograph.

7.1.2. *Temperature and humidity*

(1) Dry- and wet-bulb temperatures

Since air absorbs solar radiation very poorly there is negligible difference between the air temperature in the sun and that in adjacent shade. On the other hand, it is easily verified that a thermometer responds markedly to sunshine which falls on it. It is therefore necessary that an instrument which measures air temperature should be shielded from direct sunshine. This is achieved in the standard louvred screen, which allows for natural ventilation while providing shelter from precipitation and sunshine. The screen is painted white so as to reflect incident radiation. It should not be adjacent to a large concrete or water area nor be overshadowed by buildings or trees.

Simultaneous readings of dry-bulb and wet-bulb mercury thermometers, at a height of about 1·2 m in the screen, provide both the

† See § 7.1.2.

temperature and humidity readings of the air, as explained in § 2.3. The humidity elements, determined from the combined T, T_W' readings by reference to a humidity slide-rule or tables derived from equation (2.31), are vapour pressure, dew point and relative humidity. The wet bulb is covered by muslin to which a wick carries a constant supply of distilled water from a jar located in the screen†.

As an alternative to values of T and T_W' obtained in the screen, a psychrometer may be used in which forced ventilation of the bulbs is provided, either by whirling the instrument by hand or by drawing air past the thermometer bulbs by means of a motor-driven fan. (In this instrument a different slide-rule or set of humidity tables must be used, for reasons discussed in §§ 2.3.4 and 2.3.5—see also problem 2.4.)

(2) Maximum and minimum thermometers

Maximum and minimum temperatures in a given period are indicated by special thermometers mounted horizontally in the screen. Maximum temperature is indicated on a mercury thermometer which has a constriction in the tube just beyond the bulb. Expansion of the mercury in the bulb forces it through the constriction, while the mercury in the column is unable to flow back through the constriction when the temperature falls; the end of the column farther from the bulb thus indicates the maximum temperature. Minimum temperature is indicated on a thermometer containing alcohol. Apart from having a low freezing point, alcohol has a large coefficient of expansion; this latter allows the use of a tube wide enough to contain a small index of dumbbell shape, the index being immersed in the alcohol. When the alcohol contracts, the meniscus drags the index back towards the bulb by surface tension; when it expands, the meniscus moves farther from the bulb leaving the position of the index unchanged. The end of the index *farther* from the bulb thus indicates the lowest temperature reached in a given period.

Both maximum and minimum thermometers are reset after reading, the former by shaking the mercury in the column back through the constriction, the latter by raising the bulb till the index reaches the meniscus when it will stop. After resetting, maximum and minimum thermometers should, of course, read the same as the dry-bulb thermometer. Occasionally the minimum thermometer reads low, owing to the formation of air bubbles within the alcohol or to the adherence of some of the liquid to the tube above the meniscus. In either case the fault may be rectified by vigorous shaking directed towards the bulb end.

Grass-minimum temperature is recorded by an alcohol minimum thermometer supported on Y-pegs so that its bulb is in contact with

† If the bulb is ice-covered, the scale on the slide rule which corresponds to saturation over ice must be used (in equation (2.30) L_v becomes L_s (latent heat of sublimation)). In all cases relative humidity is calculated with respect to saturation over water.

close-cropped grass. This is the exposure implied in the use of the term 'ground frost' in official forecasts. More recently, readings of minimum temperature on prepared concrete have been added, to serve as a basis for advice on road conditions.

(3) Continuous recording

The larger pattern of thermometer screen often carries a thermograph and a hygrograph, providing continuous recordings of temperature and humidity, respectively, over daily or weekly periods. The thermograph consists of a coil of two metals, of different coefficients of expansion, which are welded together. One end of the coil being fixed to the case, temperature variations cause the other end to wind or unwind and an attached pen to move up or down a calibrated chart which is wound on a clock-driven drum. For humidity, a 'hair hygrograph' is usually employed in which expansion or contraction of a bundle of human hair under tension (about $2\frac{1}{2}\%$ expansion from 0 to 100% relative humidity) is magnified in the motion of a pen arm. These recorders are, of course, not absolute instruments but have to be set at the values indicated by the thermometers and also checked, and reset if necessary, at intervals.

(4) Soil/earth temperatures

At certain stations, daily soil-temperature readings are taken, usually at various depths down to about 20 cm below bare ground. Mercury thermometers are used, with the bulb at the appropriate depth; their stems have a right-angled bend so that the graduations are on the horizontal part of the stem.

At various larger depths, down to about 1·2 m, each thermometer is suspended by a chain within an iron tube. The thermometer is embedded in wax within an outer glass tube and can therefore be withdrawn to obtain an earth-temperature reading.

7.1.3. *Precipitation and evaporation*

Rainfall measurement is, in principle, very simple. It consists of measuring the depth to which rain has accumulated in a vessel which is free from overhead obstruction, in conditions in which there has been no splash-in from the surrounding ground or evaporation from the vessel. In practice, the rainfall depth is obtained by employing a knife-edged collecting vessel of precisely-known diameter (in the British Isles, usually 5 in), and measuring the catch in a graduated cylinder tapered at the bottom to allow accurate determination of small depths. Oversheltering is considered to be avoided if no object is nearer the gauge than twice its own height. Evaporation of the catch is prevented by its being covered by a funnel. Splash-in is avoided by having the rim of the rain gauge 1 ft above its (short grass) surround. At this height, the problem of over-exposure of the gauge—i.e. the creation of eddies, in windy conditions, which tend to carry raindrops round, rather than into, the gauge—is less than at the greater heights used in some other countries†.

Recording of rainfall may be made continuous by attaching a pen arm to a float which rises in a chamber into which rainfall, caught in the gauge, is funnelled; provision is made for the automatic siphoning of the chamber when it is full, the pen thus being returned to the bottom of the recording chart. Alternatively, a collecting chamber may be divided into two equal parts, placed centrally under a collecting funnel and tilted so that the water falls into one half only. If the chamber is balanced so that it tips over (so bringing the other half under the funnel) each time an amount corresponding to a fall of 0·2 mm, say, has been collected, and if it is arranged that an electrical circuit is made or broken at each change-over, then remote recording of the rainfall, in steps of the chosen amount, is easily arranged.

Since snowflakes are even more liable to the effects of eddies than are raindrops, a rain gauge is of little use for snow measurement. The normal method is to collect a cylinder of level snow, of known diameter, and measure its water content on melting. (Approximately, the ratio of snow depth to water-equivalent depth is about 10 to 1.)

Evaporation measurements are relatively few in number. The usual method involves the use of an evaporation tank which is nearly filled with water and sunk into the ground with its rim projecting about 6 cm. The water level is kept at about this same distance below the rim by adding or removing water as required, depending on the rates of rainfall and evaporation. Daily readings of level are made by means of a micrometer gauge and the evaporation (in mm) is deduced. Because of the exposure factor it is not yet really known, however, how these measurements compare with the rates of evaporation from natural surfaces. (The interception of cloud water by vegetation in regions affected by hill fog is a further factor which severely limits the accuracy with which the local water balance of such regions is known.)

7.1.4. *Wind*

The instrument exposure aimed at in the measurement of surface wind is an unobstructed height of 10 m in the locality concerned. What the actual height should be at any particular station is determined rather empirically; in practice, a height somewhat greater than 10 m above ground is usually required to compensate for the effects of buildings etc. on the airflow.

Wind direction is measured by a wind vane which is aligned, by the wind force acting on it, so that it points into the wind (see Plate 10). Wind speed is measured by an anemometer, which usually comprises three hemispherical cups, symmetrically disposed about a vertical axis. The difference between the drag on the concave and convex faces causes

---

† On exposed sites, shelter is sometimes provided by a circular turf wall, 1 ft high and with sloping outside wall, round the gauge and about 5 ft from it.

a rotation of the cups, and of the vertical shaft, at a speed proportional to the wind speed. In the British Meteorological Office pattern a voltage is generated by the rotation and is recorded as a continuous measure of wind speed.

7.1.5. *Clouds and visibility*

The classification of clouds in reports for synoptic meteorology is based on the cloud genera discussed in § 4.3. The heights of the bases of low clouds above ground are also reported. These reports are usually based on estimates, but when clouds at airfields are so low as to be a hazard to aircraft, height of cloud base is often measured. During the day, this involves finding the time taken by a hydrogen-filled balloon, released at the ground and rising at a known rate, to disappear in the cloud base. At night, a cloud searchlight at a distance L (1 km, say) from the observer projects vertically; if the measured angle of elevation subtended by the illuminated spot on the cloud base is E, then the cloud base is at a height $h = L \tan E$.

Visibility is the farthest distance at which a suitable object can be clearly recognized. It is an important element in aircraft operation and is also useful in air mass analysis (Chapter 8). In practice, each station has a list of visibility objects as near as possible to standard distances. Night visibility is intended to represent the same degree of atmospheric clarity as daytime visibility but presents obvious practical difficulties of estimation.

7.1.6. *Sunshine and radiation*

The difficulty of sunshine measurement is essentially one of instrument response in relation to broken overhead conditions and to occasions of low solar angle (and thus low intensity of the solar beam). On a given day it is unlikely that different types of sunshine recorder would measure the same total number of hours of sunshine: accordingly, the Campbell–Stokes type of sunshine recorder is standard at all climate and synoptic stations (at least in the British Isles).

This comprises a glass sphere resting on an adjustable support and concentric with a metal collar slotted to take the cards on which the record is made. The collar is supported in such a way that adjustment for latitude is possible. The action of the recorder depends on the charring of the card by the heat of the sun's rays focused on it by the glass sphere. The effectiveness of this simple system depends on the uniformity of the sphere's focal length, on its cleanliness and on the precise texture of the specially impregnated cards.

Although 'number of hours sunshine' is a vital index for holiday resorts, it clearly does not provide an absolute measure of solar input. Accordingly, a number of stations (all too few perhaps) determine as a matter of routine the short-wave radiation intensity on an upward-facing horizontal surface. The surface used is a blackened thermopile, i.e. a bank of thermocouples, covered and protected by a glass, crystal, or quartz dome (none of which materials transmits beyond 3 to 4 μm)†. Such an instrument is called a solarimeter; it detects both direct and diffuse solar radiation (see fig. 3.2). The latter only is measured by mounting a metal band to keep the receiving element in permanent shadow. An inverted solarimeter detects the short-wave radiation reflected by the surface.

A similar instrument which has no protecting dome is sensitive to all incident radiation and is termed a radiometer. If the sensing element is a thermopile, in the form of a flat plate, with its hot junctions facing upwards and its cold junctions downwards, and if this is suitably exposed, we have a 'net radiometer'; such an instrument is used extensively in research into local heat and energy budgets.

7.1.7. *Ship observations*

The very valuable observations provided by ships encounter special exposure difficulties. Thus a specially designed barometer is needed to minimize the effect of excessive 'pumping' of the mercury caused by motion of the ship. Wind velocity is measured by observing the state of roughness of the sea (so deducing the appropriate Beaufort force from a scale which is laid down) and observing also the direction of movement of the waves raised by the wind. Sea temperature is measured by a bucket method in which a sample is drawn from the top 30 cm or so, or by measuring the temperature of the water which is taken into the engine room from a depth of about 1 m; either of these is regarded, because of the normal mixing of the surface layers, as the sea-surface temperature.

Because of the pronounced eddying motions of air near a ship, and also because of the frequent occurrence of spray, no satisfactory method of measuring rainfall amount has yet been devised. The absence of direct rainfall measurements from sea areas limits the accuracy with which the water balance is known on a global scale.

7.2. *Upper air observations*
7.2.1. *Historical*
(1) Early soundings

Attempts to measure atmospheric conditions away from the Earth's surface began about the middle of the 18th century, the earliest recorded being that of Dr. Alexander Wilson of Glasgow who used kites to lift thermometers into the air. At almost the same time, in Philadelphia, Benjamin Franklin was carrying out a famous series of experiments on point discharge of electricity and in 1752 he also made use of a kite to investigate the electrified nature of thunder-clouds—research which

---

† The working e.m.f. is produced by having the cold junctions of the thermocouples in thermal contact with a surface which is painted white with magnesium oxide.

quickly bore fruit in the introduction of lightning-conductors for the protection of tall buildings.

The next major landmark in atmospheric exploration came just a century later when, in 1852, John Welsh, superintendent of Kew Observatory, London, made four personal balloon ascents in which he succeeded in measuring pressure, temperature and humidity to a height of nearly 7 km. Observations of these elements were made in various countries at intervals during the remainder of the 19th century, latterly by means of recording instruments which were carried aloft by kites or hydrogen-filled balloons. The results obtained were consistent in that—minor irregularities apart—temperature was found to decrease with increasing height. This was therefore regarded as a universal state of nature so that it came as a great surprise when Teisserenc de Bort, in France, using sounding balloons made of varnished paper, was able to show that, at a sufficiently high level, temperature ceased to fall with increasing height. He called the lower part of the atmosphere, which is characterized in general by a lapse of temperature, the troposphere, and the atmosphere above, where conditions are (nearly) isothermal, the stratosphere; subsequently, the term 'tropopause' was proposed for the level at which the troposphere ends.

In the early years of the present century, atmospheric soundings increased in number and were extended to various parts of the world, including sea areas. Continued use was made of kites, and great heights could be attained by attaching several of them (as many as seven were used) at intervals along the line. The latter was made of fine steel wire because of the very considerable strain involved. On some occasions, when the winds were stronger than expected, the strain proved too great and trailing steel wire, in lengths of several kilometres, was then capable of creating considerable havoc at ground level, especially near large centres of population. The use of balloons, which by this time were made of rubber, avoided these particular difficulties but had the disadvantage that recovery of the instruments and record was much less certain. Because of the large premium on the weight of instruments which were to be raised to a high level and also because of the need to restrict costs—in those days an especially important factor—the design of the instruments reached a high degree of refinement and simplicity; in the 'meteorograph' designed by W. H. Dines, for example, the record of temperature or humidity against pressure, scratched by a steel pen on a metal plate, was only about the size of a postage stamp.

The extension of temperature soundings to many parts of the world in the early years of the present century established the further unexpected feature that the lowest temperatures of all were attained near the tropopause above the equator. During this period, also, the visual tracking of so-called pilot balloons by theodolite was increasingly used to measure winds aloft. Air sampling was even possible by a device, employed by de Bort, in which, at a prearranged pressure corresponding

106

to a height of about 14 km, an evacuated glass bulb which had been carried aloft by balloon was automatically opened to admit a quantity of air and was then immediately closed. Ground analysis of the recovered sample verified that there was no difference in composition from that of surface air; this result showed that the atmosphere up to this level was well mixed—hence the appropriateness of the term troposphere (mixing sphere) for this part of the atmosphere.

Since the balloon soundings of pressure, temperature and humidity involved, at best, a delay of several days before recovery and analysis, they played no direct part in forecasting. Atmospheric soundings from aeroplanes were, however, used during World War I to obtain up-to-date information on conditions in the upper atmosphere and these became routine in many countries after about 1920 as the growing use of air transport imposed increasing demands on weather forecasting. Aircraft sounding remained the standard method of sampling the atmosphere until supplanted in 1940–45 by radiosondes of the type described in § 7.2.2.

When the isothermal nature of the lower stratosphere was established, it was generally assumed that these conditions would prevail outwards beyond the ceiling of sounding balloons to the fringe of the atmosphere. Many years elapsed before this assumption was refuted by the results of soundings by exceptionally large balloons (reaching about 40 km) and by rockets (reaching about 100 km)—see fig. 2.1, which is based on the results of such soundings. However, a similar temperature structure to that of fig. 2.1 had been inferred earlier from ground observations of various phenomena. The nature of these observations and the deductions drawn from them are worth outlining as an illustration of the power of acute inference allied to careful observation.

(2) Noctilucent clouds; sound propagation; meteors

The types of observation concerned are three in number. One of them relates to the occurrence of so-called noctilucent clouds which are sometimes seen in latitudes greater than about 50° during night hours of summer months†. These clouds are unique in that they occur at a level about five times that of the highest of ordinary clouds. Their actual heights are obtained in the following way. Simultaneous photographs of the clouds are taken at two (or more) places some considerable distance apart (200 km, say). The same identifiable cloud feature on each photograph is then measured against the star background and the height of the cloud is calculated by trigonometry, the distance between the observing points being known. The height is thus found to be in

† Although the Earth's surface and lower atmosphere are in darkness, the air at very high levels is still in sunshine during this season in higher latitudes but is too tenuous to scatter light down to Earth strongly enough to be visible. However if clouds are present, the particles are numerous enough for the light which they scatter towards Earth to be seen by an observer unless lower clouds intervene.

the range 80–85 km. For the reasons discussed in § 2.3.6 W. J. Humphreys postulated, in 1933, a much lower temperature at this level than occurs in the lower stratosphere. He did so on the assumption that the cloud particles are composed of ice. Although there has been more recent controversy as to whether they are ice or dust particles, samples collected by rocket have shown evidence of ice. Further, measured temperatures at 85 km in summer are low enough (about −110° C) for ice to exist—the corresponding saturation vapour pressure being only a small fraction of the total pressure at that level.

The second type of observation relates to what is called anomalous audibility. Before discussing it, we shall consider in a general way how temperature distribution affects sound propagation through the atmosphere. We shall not discuss the effect of wind except to say that audibility is greater in the downwind direction, provided—as is normal—wind increases with height.

Sound waves move faster through warm air than through cold, their velocity being proportional to the square root of absolute temperature. Under the normal conditions of a low-level lapse of temperature, the waves therefore bend upwards as they move outwards from a ground source and the range at which they are audible is restricted accordingly. If, on the other hand, there is a surface inversion of temperature the waves are bent back towards the ground and the range of audibility is increased. This is the explanation for the observed tendency for audibility to be greater by night than by day, and in polar latitudes (where surface inversions are common) compared to lower latitudes.

An anomalous effect is observed with large explosions. Typically, these are audible out to a range of about 100 km, are then inaudible for the next 100 km or so (zone of silence), and are then heard again for perhaps another 100 km. The first distance of 100 km corresponds to the range of the sound waves which travel near the surface. The sound heard between 200 and 300 km, say, is produced by waves which first travel outwards and at an angle from the source until they encounter a warm region, as the result of which they are bent back to Earth; more than one explosion may be heard in this outer zone, corresponding to waves which meet the warm layer at different angles and therefore penetrate it to different depths and so have different path-lengths. The waves which travel upwards from the source fairly near the zenith meet the warm region at too large an angle to be refracted back to earth, thus accounting for the zone of silence; the waves which reach the ground immediately outside the zone of silence have met the upper warm layer at just a sufficiently acute angle to be bent back to Earth†.

The existence of a warm upper region could reasonably be inferred from the existence of anomalous audibility. However, careful measure-

---

† An analogy may be noted with the passage of light from a more dense medium to one which is less dense; in such a case there is a critical angle of incidence at which total reflection just occurs at the boundary of the two media.

ments involving directional microphones and precise timing of arrival of sounds are required for accurate calculations of the paths of the sound waves and of the temperature distribution at high levels.

The third of the observations previously referred to concerns the heights of appearance and disappearance of meteors. Particles which enter the Earth's atmosphere are heated by the frictional resistance of the air and, apart from the few which arrive at the surface as meteorites, are evaporated in this way. The track of many such meteors is visible from the ground, and from synchronized observations at points a known distance apart, the heights of their appearance and disappearance may be measured and statistics compiled.

Since the level at which meteors appear and the rate at which they evaporate both depend on the number of collisions with air molecules, i.e. on ambient air density, the observations imply a particular distribution of density with height. The densities which were first calculated in this way (by Lindemann and Dobson in 1922) were about 1000 times higher than would be the case if the atmosphere were isothermal and had the same temperature as the air at the tropopause (as was then considered to be probable). The results indicated a temperature maximum at about 50 km of about the same value as at the Earth's surface—a conclusion which was not at that time accepted but which is now known to be essentially correct.

7.2.2. *The radiosonde: radar winds*

Radiosondes used for routine meteorological soundings are employed in such large numbers that their design must be compatible with cheap mass-production. In addition there is an obvious restriction in weight, usually to less than 2 kg. Such a radiosonde, therefore, is invariably the product of a neat exercise in cost-effectiveness and, consequently, is a model of simplicity.

In most types of radiosonde, pressure, temperature and humidity sensors are switched alternately into an electric circuit which generates appropriate signals suitable for transmission to base. Differences between various types lie for the most part in the precise method used to convert the mechanical motions of the sensors into transmittable signals. The construction and mode of operation of the current British Meteorological Office radiosonde (Mark 2B) are described in the following paragraphs.

The pressure sensor is an evacuated, steel, aneroid capsule. This responds almost linearly to changes in pressure, a deflection of about 2 mm resulting from a change of 1000 mb. Because of elastic hysteresis effects, all units are calibrated with a falling pressure so that the calibration is compatible with pressure changes on ascent. No compensation is made for the effect of temperature changes on the sensor's elastic modulus.

The temperature sensor is a bimetallic strip in cylindrical form,

10 mm in diameter and 16 mm long. Ordinary and invar steel are combined in such a way that the cylinder curls up with increase in temperature. Each sensor is cycled between $+60°C$ and $-75°C$, five times at least, to relieve internal strain and improve its stability.

A rectangular duct of polished aluminium shields the sensor from solar radiation. This is effective enough at the lower levels, but at high levels where reduced air densities decrease the cooling effect of forced convection† the air reaching the sensor is heated appreciably in daytime by the shield itself. In addition, the lag in the sensor's response to changes in air temperature becomes significant as densities decrease.

The humidity sensor consists of a strip of gold-beater's skin, 6·4 mm wide and 15 mm long, kept under tension. The skin stretches with increase in relative humidity, but must be seasoned for at least 30 minutes in a saturated (circulating) environment before reproducible readings result. Even then, gold-beater's skin exhibits a hysteresis effect if temporarily exposed to relative humidities below 30%. More serious than this, however, is its slow response to changes in relative humidity, particularly at low temperatures. (This lag may explain the failure of most radiosonde ascents to indicate 100% saturation in cloud-filled layers).

Each sensor is linked to the armature of an inductor, so that changes in each of the meteorological elements result in changes in the corresponding inductances. A mechanical switch, operated by a simple 3-cup windmill, acts in such a way that each inductance, in the order pressure–temperature–humidity–temperature–etc., forms part of an audio-frequency oscillator whose frequency of oscillation depends on the inductance value. This audio-frequency signal, which varies between 500 and 1000 Hz, is used to modulate a radio-frequency carrier signal (of about 27·5 MHz) which is transmitted to base.

Electrical power for the oscillator and transmitter is provided by a special light-weight battery. The battery is encased in cellulose wadding with the result that its inevitable failure at low temperatures is sufficiently delayed.

Wind speed and direction are found by radar tracking of a specially-designed wire-mesh reflector attached by cord to the balloon (the sonde, in turn, being slung beneath the reflector). Heights are found from slant range and elevation angle, while the corresponding vector winds are derived from increments in these quantities and in azimuth angle. (Radar height-finders are not as yet accurate enough for purposes of isobaric analysis‡ (see § 6.5.3), particularly in strong winds when elevations $<5°$ are common§; isobaric heights are determined by

† The sonde rises at about 7 m s⁻¹.
‡ Especially at sea where errors arise from motion of the ship.
§ Errors here may amount to 1% of indicated height: e.g. as much as 100 m at 10 km.

repeated application of equation (2.12) in virtual temperature form). Radar winds are found four times daily, while comprehensive radio-sonde ascents are made twice daily, at each station.

At present a redesigned British radiosonde (Mark 3) is undergoing evaluation. Most worthy of comment is perhaps the resistive tempera-ture sensor used instead of a bi-metal element; a 2 m length of 13·5 μm diameter tungsten wire (resistance $10^3\Omega$ at 40°C) is coiled in an open helix of length 0·25 m and mounted in two (horizontal) circular sections. Because of the effectiveness with which heat is convected away from such a device it is particularly insensitive to incident radiation, and error due to lag is minimized.

7.2.3. *Ozone measurements*
Average features of the distribution of the total amount of atmos-pheric ozone are illustrated in fig. 10.7. The variations are often large from day to day as well as seasonally. Despite the inaccessibility of the gas, it is possible to make surface measurements of the total amount held in an atmospheric column with an accuracy of a few per cent. The instrument employed is a spectrophotometer which was developed by G. M. B. Dobson. More recently, ozone sondes have been used to measure the vertical distribution of ozone.

(1) Spectrophotometer
In the spectrophotometer method, the measurement is based on the degree of absorption by ozone of certain wavelengths in sunlight. The wavelengths concerned are in the region of partial ozone absorption in the near ultra-violet (0·30 to 0·33 μm). In practice, two wavelengths are used, one strongly absorbed and the other weakly absorbed. The wavelengths are isolated and allowed to fall in rapid succession on a photomultiplier which is connected to a galvonometer. The intensity of the longer wavelength, which is less strongly absorbed than the other, is reduced by a calibrated optical wedge until the photomultiplier outputs are equal, as indicated by the galvanometer. The amount of ozone in the vertical column is obtained from the position of the wedge and the solar zenith angle at the time of observation. The calibration of the wedge involves the inferred intensities of the selected wavelengths outside the atmosphere. This is obtained from a series of observations, at various solar zenith angles, extrapolated to zero path-length through the atmosphere.

The most accurate measurements of ozone amount are those made against direct sunlight. Measurements are also made against the zenith sky, either clear or cloudy.

(2) Ozone sonde
One form of ozone sonde, used to a limited extent, is that devised by A. W. Brewer. Air is bubbled through a small electrolytic cell filled with neutral potassium iodide solution (KI). The cathode is a piece of narrow-gauge silver wire and the anode fine platinum gauze. If the

111

air contains ozone, iodine is formed in the solution, as follows:

$$2 \, KI + O_3 + H_2O = 2 \, KOH + O_2 + I_2.$$

This iodine is reduced electrolytically at the cathode:

$$I_2 + 2e = 2 \, I^-$$

and the iodine ions (I^-) are conducted to the anode where they are neutralized. A current of up to 5 μA is produced in this way, and on amplification may be converted into a suitable radiosonde signal. The entire assembly weighs about 0·6 kg and is designed to be interchangeable with the sensors in a standard Meteorological Office sonde.

(3) Rockets and satellites

Through the application of balloon techniques, the climatology of the vertical distribution of ozone is now fairly well known—up to 25 km at least. Above the ceiling of balloons, information is provided by rocket sonde or by satellite. The former is expensive in terms of quantity of data produced, each rocket ascent providing but one profile. Satellites, on the other hand, can provide information from over the entire atmosphere—or, more precisely, from two points in every orbit. An ultra-violet detector aboard the satellite monitors the rate of extinction of a selected wavelength in the direct solar beam during the moments before 'sunset', and its rate of revival after ' sunrise '; from these observations vertical profiles of ozone may be computed.

7.3. *World Weather Watch*

Advances which have been made in recent years in space technology, automatic weather stations and the like have made it possible to plan for, and to embark on in a preliminary way, an ambitious programme of global atmospheric surveillance known as World Weather Watch. This is designed to meet the direct observational requirements of the International Global Atmospheric Research Programme (GARP)† and is expected to become fully operational by about the mid-1970's.

The global observations which are visualized will obviously rest in part on the networks which are already established throughout the world. However, the required coverage cannot be met simply by extending routine techniques and existing facilities, which fall very far short of GARP requirements over the oceans, in tropical regions, and over almost the entire southern hemisphere. Evaluation is now being made of a system based largely on remote sensing of the atmosphere from satellites, either indirectly by spectral analysis of outgoing longwave radiation, or directly by interrogation of automatic weather stations in isolated regions or of sensors carried by drifting balloons. In the following paragraphs some indication is given of likely developments.

† The fundamental aim of GARP is to investigate as fully as possible the predictability of the atmosphere, based on complete knowledge of its large-scale initial structure.

PLATE 1. Small **cumulus**, with **cirrostratus** approaching from the west ahead of a weak warm front. Edinburgh, 25 July 1969.

PLATE 2. **Cumulonimbus** anvil in unstable north-westerly air stream. Edinburgh, 15 April 1969, looking south-west.

PLATE 3. **Cumulus** street, with **altocumulus** above; **cirrostratus** to the east marks a front which passed earlier. Edinburgh, 23 July 1969, looking north.

PLATE 4. **Altocumulus**, lenticular to north-east and more dense overhead. Edinburgh, 21 May 1969.

PLATE 5. Stationary wave clouds; lenticular **altocumulus** over Antarctic peninsula (65°S), seen from the west. Photograph by J. C. Farman, 19 August 1957.

PLATE 6. 'Mont Blanc fume sa pipe.' Banner cloud seen at a distance of 7 km from the south-east. Photograph J. C. Farman, August 1961.

PLATE 7. **Stratus** formed by uplift over Table Mountain, Cape Town, South Africa. 19th century photograph by G. W. Wilson.

PLATE 8. **Stratocumulus** formed by the spreading-out of early morning cumulus in subsiding maritime arctic air. Easter Ross, looking north-west, 19 May 1969.

PLATE 9. **Altostratus** veil in advance of warm front; **cumulus** formed earlier is damped down. Edinburgh, 10 May 1969.

PLATE 10. **Nimbosratus.** Persistant rain; surface wind easterly. Edinburgh 22 April 1969.

PLATE 11. Cold frontal clearance. Edge of frontal **nimbostratus**, with **altocumulus** in distance; **stratus** lingers on hills below. Ben Nevis summit, looking south-west, 26 July 1969.

PLATE 12. **Cirrocumulus** seen through passing **cumulus**. Edinburgh, 27 June 1969.

PLATE 13.   The Earth seen from a distance of 22 300 miles.
10 November 1967.   NASA photograph.

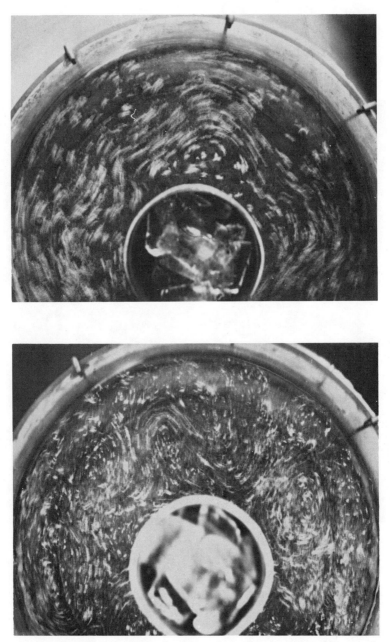

PLATES 14 and 15. Reproductions of motion in a rotating vessel, obtained by short (half to one second) time exposures by a camera rotating with the vessel. The presence of both long waves and vortices is shown.

Apart from measuring the total outgoing radiation, which is a fundamental quantity in the overall heat budget of the Earth-atmosphere system, satellites may detect the outgoing radiation intensity in various selected bands of the spectrum. For example, the atmosphere is more or less transparent to infra-red radiation in the bands $3 \cdot 4$–$4 \cdot 2$ μm and 8–12 μm, so that the intensity of the outgoing radiation within these limits clearly provides a measure of the (radiation) temperature of the underlying surface, whether it is land, sea or cloud. As cloud temperatures usually decrease with height it is easy to transform observed cloud temperatures into approximate cloud heights. Clearly, then, infra-red cloud photography can provide more information on cloud structure than is given by the present visual techniques (shown in fig. 11.2 and Plate 13). Moreover, infra-red cloud photography is as effective by night as it is by day, with obvious advantages.

It has been shown that it is possible to measure temperature profiles down to the levels of the highest clouds by observing those wavelength bands in which there is absorption, and therefore re-emission, by CO_2†. However, sufficient accuracy has yet to be attained. The feasibility of using radiation measurements within the water vapour absorption bands† to determine the vertical distribution of water vapour between 5 and 10 km has likewise been demonstrated with some success.

The most promising development in the direct sensing of atmospheric structure for global application involves the use of large numbers of constant-volume balloons, floating at several selected constant density levels and tracked by one or two satellites in polar orbits. This project, known as GHOST (Global Horizontal Sounding Technique) allows mean vector winds to be determined for each balloon, over given periods, while pressure, temperature and humidity (measured by attached sensors) are transmitted to one or other satellite on interrogation. In a preliminary experiment, carried out in the southern hemisphere, one particular balloon stayed at around the 200 mb level for 234 days, and completed 20 circuits of the Earth before being lost. However others of the balloons, at lower levels, failed after only a few days, owing to icing. Provided that the icing problem (and others) can be overcome, it is likely that this technique will form an important part of World Weather Watch.

For surface observations in remote regions the use of automatic weather stations is likely soon to become commonplace. Several automatic observing systems are at the prototype stage and are proving to be practical in operation. One such British system can be interrogated over ordinary telephone lines, but for stations in the most remote land areas and for those on tethered buoys at sea, satellite interrogation is intended, with subsequent retransmission of the collected observations to central data stations for decoding and analysis.

† See fig. 3.3.

The successful implementation of World Weather Watch will obviously involve international effort and collaboration on a scale greater than ever before. Meteorologists consider the project essential for the better understanding and prediction of the atmosphere.

7.4. *Experiments in observation and interpretation*

We end this chapter with a number of suggestions of possible experiments, most of them involving very simple apparatus. For some experiments the possession of instruments which provide a record of pressure, temperature, humidity and rainfall is assumed; in the last case it is not difficult to construct a satisfactory remote-recording instrument (see § 7.1.3). Where such (recording) instruments are used they should be time-marked at appropriate intervals during each day or week and allowance made for any errors of indicated times.

An accurate and easy measurement of temperature and humidity is provided by the whirling psychrometer, an instrument which is easily available commercially—as are inexpensive solarimeters and hand anemometers and also sensitive (Sheppard-type) cup anemometers. Satisfactory thermometer mounts, for 'profile' experiments, can be made from any suitable material, in tubular form and painted white. (Join a 25 cm length of 2 cm diameter tubing, on which the stem of the thermometer rests, (concentrically) to a 10 cm length of 4 cm diameter tubing which shields the thermometer bulb without being in contact with it.)

7.4.1. *Pressure*

(1) From a single daily reading of pressure, or from a midday reading taken from a barograph, correlate pressure against daily rainfall amount.

(2) From continuous records of rainfall and pressure find how rainfall amount is related to pressure tendency.

(3) Using only days of small pressure range, as judged from a series of barograph traces (perhaps 20% of the total), determine whether there is a systematic diurnal variation of pressure. Test if possible the statistical reliability of the result, and find an explanation for it. (It will suffice to read off pressure values at 2-hour intervals, starting at 00 h. Any difference between hours 00 and 24 for the average day should be removed by smoothing it out over the entire day, as such a trend represents a non-cyclic type of change.)

7.4.2. *Temperature and humidity*

(1) Make a series of comparisons of dry-bulb and wet-bulb temperature readings (and of derived values of dew point and relative humidity) obtained from a thermometer screen and from a whirling psychrometer in different conditions. Determine whether systematic differences between the observations occur. Relate such differences as do occur to the prevailing weather conditions.

114

(2) Measure vertical profiles of temperature on a ' ladder ' (a simple support for thermometer mounts) at, say, the surface, 10 cm, 20 cm, 50 cm, 1 m, and 2 m, and in the soil at 5 cm, 10 cm and 20 cm (standard depths); find by hourly readings the way in which each profile changes during a sunny day. Determine the profile of minimum overnight air temperature and of maximum daytime air temperature. Graph and discuss the results in relation to amount of cloud, wind speed, type and condition of surface, etc.

(3) During clear weather, place a number of minimum thermometers on short grass close to and at 1 m intervals from, (a) the roots of a hedge, (b) the trunk of a tree, (c) a stone wall, (d) an open-work fence. Sketch the isotherms implied by the recorded minimum temperatures.

(4) Construct a temperature-humidity mast on which a number of thermocouples are mounted, one ' dry ' and one ' wet ' at each of the six levels suggested in experiment (2). Each pair of thermocouples is shielded from direct radiation by a white cylinder; this is aligned with the wind to ensure maximum ventilation. (Several loops of thread or fine wool wound round a thermocouple junction and kept moist with distilled water constitutes a wet thermocouple.) The e.m.f. generated between any one of the mast junctions and a standard junction is a measure of this temperature difference; the temperature of the standard junction, measured with a mercury-in-glass thermometer, should be recorded at suitable intervals. Dry- and wet-bulb temperatures may thus be determined at each level in turn with the aid of a thermocouple switch and a simple potentiometer bridge.

Obtain profiles of temperature, vapour pressure, dew point and relative humidity, (a) over short grass, (b) over tar-macadam, (c) over concrete, (d) in long grass or a field crop, (e) to windward and leeward of a lake. Relate the observations to the nature of the underlying surface and to overhead conditions.

7.4.3. *Evaporation and rainfall*

(1) Determine the hour-by-hour evaporation rate from a water surface during a cloudless day. To do this at all accurately some form of amplification of the change in water depth is required; the following mechanical technique is perhaps the simplest.

Set a large water bath (1 m or so in diameter) solidly into the ground and fill it with clean water, as described in § 7.1.3. Obtain a light, tapered pointer about 1·5 m long and attach its heavy end to a float weighted to displace about 2·5 litres of water. Fix the pointer to the edge of the water bath in such a way that it is free to pivot about a horizontal axis; let this axis be at such a distance† from the centre of the float that vertical motion of the float is amplified 10 : 1 in the motion of the far end of the pointer. Solidly mount a millimetre rule close to but not quite touching the end of the pointer. Fit a length of

† In this event 13·6 cm.

narrow-gauge piano-wire as a short extension to the pointer, to facilitate accurate reading of the scale.

Read the level of the water in the evaporation bath every hour (or half hour) to the nearest 0·05 mm, i.e. to the nearest 0·5 mm on the scale; make a plot of this and of the derived evaporation rates, expressed in mm per day. Plot also the corresponding values of air temperature, relative humidity, wind speed, net radiation (if available) and water temperature, the last obtained from a mercury-in-glass thermometer attached beneath a small tethered float.

(2) Obtain about twenty identical bottles or cans and a graduated measuring cylinder. Place the containers either (a) around a building, (b) through a wood, (c) across a quadrangle, (d) across the line of a hedge, or (e) in any array on a region of open ground.

Compare measured catches after significant overnight rainfall and relate to wind direction. Present the results in map form, showing isohyets—i.e. lines of equal rainfall amount. Alternatively, place an identical amount of water in each container early in the day and measure later in the day the amount that remains. Plot the results to show lines of equal evaporation, after correction for rainfall if any.

(3) From one year's daily records, chart rainfall amount against wind direction. Compare and contrast this with the corresponding ' wind rose ' (see below).

7.4.4. *Wind*

(1) From a month's observations of wind speed and direction at a particular hour of the day construct a wind rose; i.e. plot on a polar diagram, in histogram form, (a) the total number of days with wind from the north, north-east, east, south-east, etc., (b) the respective numbers of days with· winds over, say, 10 kt (shade this area lightly), and over 20 kt (heavy shading). Do this for each month of a year and for the year as a whole.

Obtain for each month the departure of its mean temperature (calculated from $\frac{1}{2}(\text{min} + \text{max})$ for each day) from the long-term monthly mean and indicate this departure beside each wind rose. Show that there is a correlation between the sign of the temperature anomaly and the mode of departure of the wind rose from its usual shape (taken as indicated by the rose for the entire year).

(2) With a hand anemometer investigate, on a fairly windy day, the degree and extent of the shelter provided by (a) a wall, (b) a hedge, (c) railings, (d) a wood and (e) a shed. For (a), (b) and (c), do this at several levels, one at least being above the level of the obstacle itself. Take average wind-speed values, as judged by eye from the instrument dial, over equal periods of not less than 10 s.

(3) Mount three sensitive cup anemometers on a suitably constructed mast at heights of 20 cm, 50 cm and 2·0 m over, for example, (a) short

grass, (b) concrete, (c) long grass, (d) various field crops, (e) waste-ground, (f) a ploughed field, (g) sand and (h) the extreme end of a flat spit protruding into a lake. The upwind fetch over each surface must not be less than 80 m. By measuring the mean winds over periods, preferably of not less than ten minutes, at the three levels, determine the wind profile over each surface, and find its roughness length (see Chapter 9). (In (c) and (d) it will be necessary to adjust the zero of the height scale, by an amount d, to allow for the depth of the vegetation, h. As a first estimate†, $d = 0.75\ h$, but this fraction depends on the ' density ' of the vegetation.)

(4) Estimate the wind velocity at various tropospheric levels by observing the horizontal progression of appropriate clouds (see §§ 4.2 and 4.3.). A stop-watch is required, together with a marker and a suitable object of known elevation, to act as a foresight; e.g. a telegraph pole, a church spire, or a chimney head. Let h be the height of the foresight above the observer's eye-level, and let H be the estimated height of the cloud, or part of the cloud, in question. With the chosen cloud positioned behind the top of the foresight, drop the marker and start the stop-watch. Move so that the cloud remains ' attached ' to the foresight, until a distance d metres, of about $h/3$, has been covered; then stop the watch. If time t seconds has elapsed, then the wind speed at the level in question is $(d/t) \times (H/h)$ m s^{-1} and its direction is given by the vector from the observer to the dropped marker.

7.4.5. Radiation

(1) Make the following use of a simple uncalibrated solarimeter. Obtain the albedo of various surfaces such as grass, soil, sand, concrete, water, ice, snow, etc. (Albedo is the ratio of the output of a solari-meter when facing downwards to its output when facing upwards, expressed as a percentage.)

(2) Measure the ratio of diffuse to total solar radiation at different times of day and under different overhead conditions. Show that this ratio increases in the presence of cloud and with decrease in solar angle. (Simple means may be used to shield the solarimeter from direct solar radiation.)

(3) Plot the short-wave radiation intensity measured at frequent intervals during a traverse through a wood.

(4) In an area of open ground (well away from buildings, trees, etc.) make a series of observations with the solarimeter facing horizontally towards each of, say, 16 compass points in turn. Do this at different times of the day and under various (preferably steady.) overhead conditions. Plot the readings on a polar diagram.

---

† d is known as the zero plane displacement. When required, it is included in equation (9.10), thus: $u(z) = u_*/k. \ln\ ((z-d)/z_0)$. Its value is found empiri-cally as that which gives a linear plot of $\ln\ (z-d)$ against $u\ (z)$.

7.4.6. *Topographical influences*

(1) With a hand anemometer and a whirling psychrometer, investigate the behaviour of wind speed, air temperature, dew point and relative humidity during each of the following traverses: (*a*) up, over and down a hill of at least 100 m in height, in both a downwind and a cross-wind direction; (*b*) round a lake of at least 500 m diameter; (*c*) across a steep-sided, sloping valley in the early morning or late evening, preferably during clear overhead conditions.

(2) Use toy balloons, hydrogen-filled and Plasticine-weighted, as no-lift tracers to investigate natural airflow in valleys and in the vicinity of hills, ridges and escarpments. It is possible to detect a return eddy in the lee of some escarpments with this technique. (Much can also be learned about the nature of airflow in the vicinity of buildings.)

BEFORE a reasoned prediction of the future distribution of the weather elements over a particular area can be made the initial distribution must be known, of necessity over a larger area, while the implications of broad-scale features or patterns in this distribution must be adequately understood. The simultaneous study of the weather elements over a large area at a given time is termed synoptic meteorology. The chief tools of the synoptic meteorologist are the charts prepared for sea level and for selected upper levels. These synoptic charts, or weather maps, provide him with a bird's-eye view of existing weather conditions at each chosen level. Such a technique of multiple representation in two dimensions of what is usually a complex three-dimensional atmospheric state is simple enough in effect to permit of a rapid analysis by man or machine, and the subsequent production of a forecast.

In this chapter we concentrate much of our attention on the surface weather map, perhaps unduly so from the viewpoint of the professional forecaster; our justification is its familiarity through the channels of television and the press.

8.1. *The surface weather map: an introduction*

Surface weather maps, such as those in figs. 8.1 and 8.2, show weather observations made at a specific time at a large number of land stations, and on ships at sea (some of them Weather Ships). Each set of observations appears at its appropriate position on the map in a simple, short-hand form.

8.1.1. *The plotting code*

The full plotting code for surface observations allows for total amount of cloud cover; wind speed and direction; visibility; present and past weather; pressure and pressure tendency; temperature and dew point; and also height, amount and type of low, medium and high cloud: knowledge of each of these weather elements at every station is required before a confident synoptic analysis can be made. However, on charts such as figs. 8.1 and 8.2, which are merely duplicates of such synoptic analyses prepared on a much reduced scale for ease of circulation and future reference, only a few weather elements need be or indeed can be plotted at each station: in consequence a reduced and somewhat simplified plotting code is used, as follows.

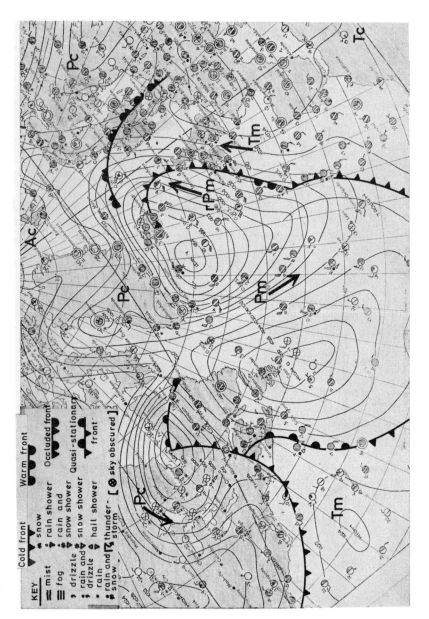

KEY

| Cold front ▼ | Warm front ● |
| Occluded front | Quasi-stationary front |

mist ≡   snow *
fog ≡   rain shower ▽
drizzle ,   rain and snow shower
rain and drizzle   snow shower
rain •   hail shower △
rain and snow   thunder-storm [⊗ sky obscured]

Fig. 8.1. Surface weather map for 1200 G.M.T. 25 January 1969.

120

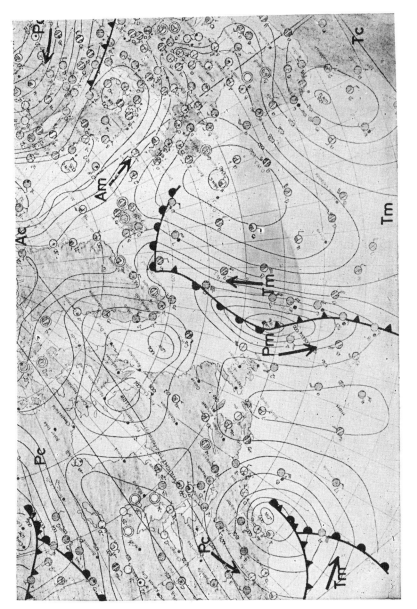

Fig. 8.2. Surface weather map for 1200 G.M.T. 9 February 1969.

121

Each station is represented by a position circle, the shaded part of which indicates the proportion of sky covered by cloud. Attached to each position circle is a feathered arrow giving wind direction and speed, each full feather representing 10 kt: a solid triangular flag represents 50 kt, while a calm is indicated by a further circle outside and concentric to the position circle. The figures plotted to the (upper) left of each station circle are surface air temperatures (i.e. at screen level) in °C, while symbols plotted below each temperature reading indicate any significant weather phenomenon at the station—e.g. snow, fog, thunder-storm, etc. The symbols used for ' present weather ' are listed with fig. 8.1.

The mean-sea-level air pressure appropriate to a particular station may, if required, be interpolated from the pressure values indicated on neighbouring isobars—the positions of which are themselves determined in the original analysis by interpolation between plotted values at neighbouring stations.

8.1.2. *Pressure systems and features*

A glance at figs. 8.1 and 8.2 is enough to reveal the existence of large-scale transient pressure systems in the atmosphere—regions of relatively low pressure being interwoven with regions of relatively high pressure. Further inspection shows that the observed wind (at a height of 10 metres) blows round low pressure centres (depressions) in a counter-clockwise sense and round high pressure centres (anticyclones) in a clockwise sense, in qualitative agreement at least with the rules of geostrophic balance introduced in Chapter 6. Moreover, the observed 10-metre wind generally inclines across the isobars from high pressure towards low pressure, a situation brought about (as indicated in fig. 6.8), by the influence of surface friction.

The primary effect of surface friction, namely that the observed wind (V) is in general less than the corresponding geostrophic wind (V_G), is easily verified at any chosen station by deriving V_G from the local isobar spacing (as in Problem 6.4): for example, at Ocean Weather Ship D (44° N, 41° W) $V_G = 30$ kt and $V = 25$ kt in fig. 8.1; while $V_G = 75$ kt and $V = 45$ kt in fig. 8.2.

Typical examples of common pressure features occur in figs. 8.1 and 8.2. A ridge of high pressure extends from ' High C ', off the west coast of Ireland (fig. 8.2), south-westwards over the Azores; while in fig. 8.1 a marked trough of low pressure extends south-south-westwards over the same region from the deep Icelandic ' Low Y '. Note that High C ' ridges ' towards Iceland and also towards the Alps and Pyrenees; while there is evidence of ' troughing ' of Low Y towards the north-east and the north-west. The terms ridge and trough may also be applied, as appropriate, to elongated regions of high or low pressure.

The region of weak and somewhat indeterminate flow between two systems of high and of low pressure, alternately disposed, is known as a col. There is a well-defined col between two deep depressions and two well-developed anticyclones—a neutral col—just off the coast of Labrador in fig. 8.1. The col at a similar position in fig. 8.2 is 'low-pressure dominated' whilst the col north-east of Bermuda, again in fig. 8.1, is 'high-pressure dominated'.

8.1.3. *Air masses*

The general circulation of the atmosphere tends to produce quantities of air of vast horizontal extent in which the physical properties (notably temperature and humidity) at any particular level are approximately uniform. Such a body of air is termed an air mass.

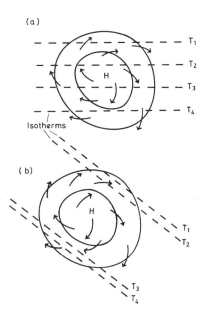

Fig. 8.3. Within an anticyclone, divergent flow weakens existing temperature gradients, (a), which concentrate around its periphery, much as indicated in (b).

The homogeneous properties of air masses are acquired in so-called source regions, characterized by uniformity of temperature and humidity. Thus snow-covered polar regions, tropical seas, and extensive desert areas all act as important source regions.

The effectiveness of a source region is greatly enhanced if it is also a region in which large anticyclones predominate. Air moves relatively

123

slowly within such pressure systems and thus has sufficient time to be influenced by underlying surface conditions and to attain equilibrium with them. Also, air continuously diverges from the central regions of an anticyclone, owing to surface friction. This ensures that a steady supply of air is available for export at the boundaries of the region, i.e. that the source is effectively continuous. Further, the horizontal divergence of the flow acts in such a way that any horizontal temperature gradients already existing in or somehow introduced to the region are progressively weakened and eliminated. As shown in fig. 8.3, such gradients concentrate at the periphery of an anticyclone.

An air mass is classified as either polar ' P ' or tropical ' T ', depending on the latitude of its source region, and as either continental ' c ' or maritime ' m '. For example, the classification ' Pc ' refers to a polar air mass over land, such an air mass having a low water content. However, a Pc air mass moistened by maritime travel is re-classified ' Pm ', i.e. polar maritime. Polar air taking part in a direct outburst towards middle latitudes is frequently classified ' A ', i.e. arctic (or antarctic).

Air mass types are indicated in figs. 8.1 and 8.2, on the lines of the preceding classification; where an air mass is clearly in motion this is indicated by an accompanying arrow. Of note in fig. 8.1 is a distinct outbreak of Pc air from Greenland into the central North Atlantic, where it is modified to Pm. Some of this modified air is returning polewards (to the west of the British Isles) and as such is designated rPm, i.e. returning polar maritime. The British Isles are immersed in a mild Tm airstream; whilst the Great Lakes are experiencing the other extreme, namely a severe outbreak of Pc air from the north-west.

8.1.4. *Fronts*

Both surface and upper-air observations show that the transition from one air mass to another tends to occur more or less discontinuously rather than gradually. It is observed that adjacent air masses are separated by a zone of transition (a frontal zone) some 100 km in horizontal extent. The colder (more dense) air mass invariably extends as a wedge

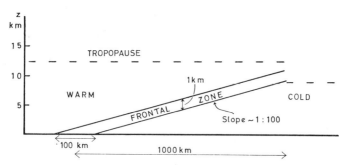

Fig. 8.4. The frontal zone between two air masses (not to scale).

under the warmer (less dense) air mass, as explained in § 8.3.1, so that frontal zones always incline towards the colder air mass—to such an extent in fact that they usually lie at somewhat less than 1° from the horizontal, i.e. at a slope of about 1 : 100.

It follows that the vertical extent of a frontal zone must be of the order of 1 km only, as indicated in fig. 8.4. This small but finite thickness is brought about by slow turbulent mixing of the adjacent air masses, turbulent mixing generated in the first instance by their relative motion (see fig. 8.8).

It is common practice to treat a frontal zone as though it were a true mathematical surface of separation between two air masses, i.e. a ' frontal surface '. A ' front ' is then by definition the line of intersection of a frontal surface with the surface of the Earth, or with any other appropriate horizontal plane.

In general, fronts are in motion. If the passage of a front causes warm air to replace cold air, the front concerned is called a warm front; whereas a cold front causes cold air to replace warm air. The conventional symbols used to represent warm fronts and cold fronts are included in the key to fig. 8.1, together with the various combinations of those symbols used to indicate quasi-stationary and occluded fronts (for the latter see fig. 8.11).

Referring back to figs. 8.1 and 8.2 we see that when isobars intersect a front they do so along a trough of low pressure, more or less well marked. (Since a trough may be non-frontal the converse is not true.) The frictional convergence towards a front which this implies has the

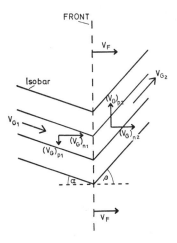

Fig. 8.5. As a first approximation the rate of progress of a front, V_F, is equal to either of the geostrophic wind components $(V_G)_{n1}$ or $(V_G)_{n2}$ (continuity at the front requires that $V_{G1} \cos \alpha = V_{G2} \cos \beta$).

125

effect of sharpening the temperature difference across it—in contrast to the weakening effect produced by anticyclonic divergence.

The speed at which a front moves, V_F, is related to the geostrophic wind component (in either air mass) normal to the front, $(V_G)_n$ in fig. 8.5. To a first approximation $V_F = (V_G)_n$; however, this equality assumes that geostrophic flow conditions exist in the immediate vicinity of fronts and this is certainly untrue since both horizontal and vertical accelerations occur there. Although $(V_G)_n$ provides a good estimate of V_F for cold fronts it is found that warm fronts travel at only about 70% of $(V_G)_n$.

In fig. 8.5 the different wind components $(V_G)_p$ parallel to the front indicate the relative motion of the neighbouring air masses. In §8.3.2 this relative motion is related to the slope of the frontal surface and to the difference in temperature between the two air masses.

8.2. Air mass characteristics

8.2.1. Classification

In §8.1.3 the basic classification of air masses in terms of source region is introduced; more details are given in the following table.

Air mass	Source region(s)	Properties at source
Polar maritime (Pm)	Oceans in latitudes greater than 50° (approx.)	Cool, rather moist; unstable
Polar continental (Pc)	Continents in vicinity of Arctic Circle; Antarctica	Cold, dry; stable
Arctic or Antarctic (A)	The Arctic Basin and (central) Antarctica in winter	Very cold and dry; very stable
Tropical maritime (Tm)	Sub-tropical oceans	Warm; moist, and rather unstable near surface, dry and stable above
Tropical continental (Tc)	Deserts in low latitudes: primarily the Sahara and Australian deserts, but also south-west U.S.A. and Mexico in summer	Hot and dry; unstable

Table 8.1.   Air mass classification.

Note that the properties of an air mass in its source region imply that it is in equilibrium with the region.

8.2.2. *Modifications*

When an air mass moves away from its source region it is no longer in equilibrium with its surroundings; thermal and dynamical influences, amongst others, progressively modify its initial properties, and may sooner or later necessitate its re-classification.

Thermal modification occurs when the temperature of the underlying surface differs from that of the source region. For example, a Pc air mass moving equatorwards is in general colder than the surface over which it passes, so that it is warmed from below and tends to become unstable. In contrast, any tropical air mass moving polewards is subjected to surface cooling and in consequence is rendered increasingly stable in its lower layers.

Thermal modification of an air mass is usually accompanied by a change in its moisture content, or at least by a change in its relative humidity. The nature of such humidity modifications clearly depends on the availability or otherwise of moisture and on whether the air mass is undergoing surface heating or surface cooling. The moisture content of Pc air moving equatorwards over land increases only slowly, while its relative humidity drops progressively to around 20%. The same air moving equatorwards over the sea exhibits a rapid increase in moisture content owing to evaporation which, with simultaneous surface heating, soon converts it into Pm air with a relative humidity maintained at around 70%. The progressive cooling of Tm air moving polewards (over land or sea) eventually raises its relative humidity to 100%, with the production of fog; and, when cooling is continued beyond this, drizzle results, with a corresponding progressive decrease in the absolute moisture content of the air.

The main dynamical influence on an air mass is the slow vertical motion to which it is subject if caught up in the circulation of a cyclone or anticyclone. When the curvature of the flow is cyclonic there is surface convergence combined with upward motion and a decrease of stability and increase of relative humidity of the tropospheric air; when it is anticyclonic each of these is reversed (see §8.7.1 and §5.4). The curvature of the isobars is a good indication of the sign and magnitude of the curvature of the flow though these curvatures are (nearly) identical only if the pressure features are stationary. Dynamical modification is often important: for example, the showeriness of a Pm air mass changes very significantly with quite minor changes of isobaric curvature from 'straight' to cyclonic or anticyclonic.

8.2.3. *Air masses over the British Isles*

Figure 8.6 indicates the most common tracks followed by the principal air mass types on their approach to the British Isles, while Table 8.2 lists their main characteristics over the British Isles.

Air mass	Characteristics
Polar maritime (Pm)	Cold in winter, cool in summer. Unstable. Visibility good. Cumulus and cumulonimbus clouds. Bright intervals and showers. Clouds and showers die out inland at night and in winter.
Returning polar maritime (rPm)	Normal temperatures. Stability variable (increases with distance travelled south by the Pm air before returning northwards). Visibility moderate. Cloud type various: often stratocumulus but cumulonimbus and showers (perhaps with thunder) may occur.
Polar continental (Pc)	Cold or very cold in winter, especially in south-east England; warm in summer except on east coast. Unstable, especially in the lower layers, except near windward coasts in summer. Rather hazy. Cloud type usually stratocumulus, but cumulus and cumulonimbus develop on the longer sea crossings to the Scottish east coast. Occasional sleet or snow flurries in winter except to west of Welsh and English mountains and Scottish Highlands where skies are often clear. In summer all inland districts enjoy clear skies and sunshine but sea fog (haar) blankets the east coast except, at times, in mid-afternoon.
Arctic continental (Ac)	Very cold or severe in winter, cold in spring and early summer. (This air mass is rare in high summer; when it does occur it is almost indistinguishable from Pc and may be warm.) Unstable at low levels. Visibility good. Cloud and weather similar to Pc at all seasons, although in winter and spring precipitation is more likely to be of snow.
Arctic maritime (Am)	Cold at all seasons; often severe in winter, especially in north Scotland. Unstable or very unstable. Exceptional visibility. Cloud types and behaviour resemble Pm, but showers are of sleet, snow or hail in winter and spring. Leeward districts experience long periods of clear sky. Night frosts severe in winter, when frost may continue all day; frost possible even in summer.
Tropical maritime (Tm)	Mild or very mild in winter, warm and humid in summer. Stable. Visibility poor, perhaps moderate in summer. Dull skies and drizzle common in west and southwest, with hill fog, but clearing to the east. Föhn effect common in lee of Scottish Highlands. Mists and fogs inland in winter and at night, when wind is light and sky clears. Some sea and coastal fog in the English Channel and Western approaches in spring and summer.
Tropical continental (Tc)	Warm or very warm in summer, mild or very mild in early winter and spring. Unstable in the upper layers; stable below. Hazy. Cloud amount variable. Stratus cloud types in winter, but in summer cumulus and occasionally altocumulus castellanus as an indicator of potential instability.

Table 8.2.   Characteristics of principal air masses over the British Isles (mainly after Lamb).

Air masses which arrive in the British Isles by tracks intermediate to those indicated in fig. 8.6 have correspondingly intermediate characteristics. In all cases due allowance must be made for dynamical (cyclonic or anticyclonic) influences on air masses.

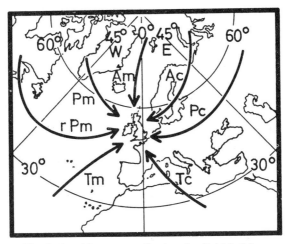

Fig. 8.6. Air masses affecting the British Isles.

8.3. *Frontal characteristics*

8.3.1. *The stability of a frontal surface*

Consider the succession of events indicated in fig. 8.7. This figure shows a laboratory-scale experiment in which two fluids of different densities, contained within the same transparent tank, are initially separated by a thin vertical barrier. On removal of this barrier, stability is not attained until the lighter fluid completely overlies the heavier and

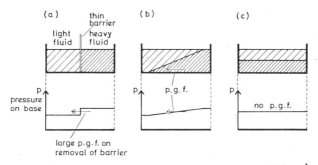

Fig. 8.7. Stability of fluids of different density in contact (laboratory scale); (a), unstable when barrier removed: (b), unstable: (c), stable (p uniform).

their surface or zone of separation is horizontal. Only then, as in stage (*c*), is the horizontal pressure gradient force (p.g.f.) near the base of the tank identically zero. It is easily demonstrated, as follows, that on an atmospheric scale the intermediate stage (*b*) is one of possible stability, owing to the existence and nature of the Coriolis force (C.f.) on a rotating earth (see Chapter 6).

Let us assume, for simplicity, adjacent warm and cold air masses with no general pressure gradient (i.e. no isobars or wind) in the warm air, as

129

shown in fig. 8.8. *a*. Just as in the laboratory-scale experiment, static pressure considerations imply that there is a p.g.f. within the denser air mass. In order that fig. 8.8 *a* should represent an equilibrium state of the atmosphere, the cold air mass must be moving bodily with respect to the warm air mass with the appropriate velocity V for which the C.f., equal to $2\Omega \sin \phi$. V, or fV, and acting at right angles to and to the right of V, is equal and opposite to the existing p.g.f. Such balanced (geostrophic) motion is indicated in fig. 8.8 *b*. Any general wind is added vectorially in both air masses, but their relative motion, i.e. the frontal wind shear, remains unchanged, as shown in fig. 8.8 *c*.

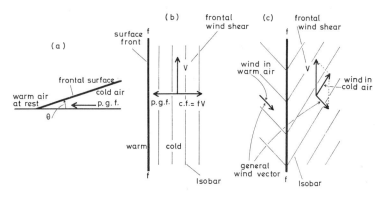

Fig. 8.8. Stability of a sloping frontal surface in the atmosphere. (a), vertical cross-section: (b), frontal wind shear required for balanced (geostrophic) conditions: (c), inclusion of a general wind vector does not affect the required relative motion.

It may be inferred from fig. 8.8 that for a particular frontal surface to be stable there must be a discontinuity of p.g.f. and an associated horizontal wind shear, each of the precise amount necessary to maintain the frontal slope at a fixed value. The equilibrium relationship between frontal slope, frontal wind shear and the p.g.f. discontinuity may be derived as follows.

8.3.2. *Equilibrium slope of a frontal surface.*

Consider neighbouring cold and warm air masses. Let the inclined frontal zone between them be replaced by a truly mathematical surface of separation, so that it appears as the line (FS) on the cross-sectional diagram, fig. 8.9 *a*. Choose a right-handed system of coordinates with x horizontal at right angles to the front and positive towards the cold air, y horizontal along the front and z vertically upwards, as indicated by fig. 8.9 *b* on which the line f F f represents the surface front. Construct the infinitesimal rectangle ABCD on the frontal surface (FS, fig. 8.9 *a*) in such a way that the frontal slope, $\tan \theta$, is given by dz/dx. Let the density of the cold air be ρ_1 and the density of the warm air be ρ_2.

130

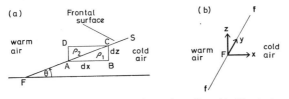

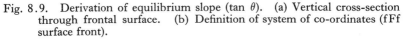

Fig. 8.9. Derivation of equilibrium slope (tan θ). (a) Vertical cross-section through frontal surface. (b) Definition of system of co-ordinates (fFf surface front).

The frontal slope is related to the difference in the pressure gradients $(dp/dx)_1$, and $(dp/dx)_2$ in the respective air masses. To find this relationship we first write

$$\left(\frac{dp}{dx}\right)_1 = \frac{p_B - p_A}{AB} \tag{8.1}$$

and

$$\left(\frac{dp}{dx}\right)_2 = \frac{p_C - p_D}{DC}. \tag{8.2}$$

It may be deduced from equation (2.7) that

$$p_B = p_C + \rho_1 g \, dz$$

and

$$p_D = p_A - \rho_2 g \, dz,$$

so that equations (8.1) and (8.2) may be written as

$$\left(\frac{dp}{dx}\right)_1 = \frac{p_C - p_A + \rho_1 g \, dz}{dx} \tag{8.3}$$

and

$$\left(\frac{dp}{dx}\right)_2 = \frac{p_C - p_A + \rho_2 g \, dz}{dx}, \tag{8.4}$$

where dx replaces both AB and DC. On subtracting equation (8.4) from equation (8.3) and rearranging:

$$\frac{dz}{dx} = \tan\theta = \frac{\left(\dfrac{dp}{dx}\right)_1 - \left(\dfrac{dp}{dx}\right)_2}{g(\rho_1 - \rho_2)}. \tag{8.5}$$

The condition that equation (8.5) should specify an equilibrium frontal slope must be that balanced flow conditions exist on both sides of the frontal surface. If we assume for simplicity that this balanced flow is geostrophic†—i.e. that it is balanced under the action of p.g.f. and C.f. only—we may write

$$\frac{1}{\rho_1}\left(\frac{dp}{dx}\right)_1 = f v_1 \quad \text{and} \quad \frac{1}{\rho_2}\left(\frac{dp}{dx}\right)_2 = f v_2,$$

† This assumption is at best an approximation because accelerations do, in fact, occur in the region of a front.

131

where v_1 and v_2 are the respective flow components *along* the front in the positive y direction. Application of geostrophic flow conditions to equation (8.5) gives

$$\tan \theta = \frac{f(\rho_1 v_1 - \rho_2 v_2)}{g(\rho_1 - \rho_2)}. \tag{8.6}$$

This equation may be translated into an approximate temperature difference form by writing $\rho_1 = p_1/R_1 T_1$ and $\rho_2 = p_2/R_2 T_2$ and observing (i) that $p_1 = p_2$ (the discontinuity at the front is in dp/dx not in p), (ii) that R_1 may be equated to R_2 with negligible error, and (iii) that T_1 and T_2, being in K, may reasonably be replaced by their mean value T, provided their difference is not of explicit importance. It may thus be shown that

$$\tan \theta = \frac{fT}{g} \left(\frac{v_1 - v_2}{T_2 - T_1} \right). \tag{8.7}$$

That is, the slope of a frontal zone is directly proportional to the geostrophic wind shear across the front and inversely proportional to the temperature contrast between the two air masses.

Although for a given temperature contrast any number of suitable pairs of values of $\tan \theta$ and $(v_1 - v_2)$ may satisfy this equation, it is found in practice that $\tan \theta$ varies little from one front (of a given type) to another (of the same type). It follows that frontal wind shear and air mass temperature contrast are nearly proportional to each other; i.e. the greater the temperature contrast across a particular front the larger is the wind shear likely to be observed across it.

Equation (8.7) consistently underestimates $\tan \theta$ for cold fronts and overestimates $\tan \theta$ for warm fronts (usual frontal slopes being about 1 : 60 and about 1 : 150, respectively). Considering that geostrophic flow conditions are assumed in its derivation this is not too surprising (in view of the remarks which conclude § 6.2.2). It is of some significance, however, that these inconsistencies should be systematic—in the sense that different departures from geostrophic flow conditions are implied for warm fronts than for cold fronts. This points to inherent differences in the dynamics of warm and cold fronts.

8.3.3. *Frontal structure*

' No idealized frontal model can adequately represent an individual front '—J. S. Sawyer. While this quotation serves as a warning of the complexity of fronts, it does not necessarily imply that a simple frontal model is of no value in an elementary treatment of frontal structure. Such a model is presented in fig. 8.10 which shows an idealized warm front in vertical cross-section. A normal sequence of clouds associated with the approach and passage of a marked warm front is indicated: cumulonimbus and cumulus cloud in the cold (Pm) air mass well ahead of the front, perhaps dying out or flattening into stratocumulus with its nearer approach; isolated cirrus in the warm air above the front, increasing and merging into cirrostratus through which solar or lunar haloes may be visible; thickening and lowering of this cloud in the warm air, first to altostratus and then to nimbostratus, with nearer approach of the surface front; behind the front mainly stratus and stratocumulus. Showery precipitation in the deep cold air well ahead of the front is followed by a fair interval, then by almost continuous rain ahead of the surface front and by intermittent drizzle behind it. (It is not, however, uncommon for the rain to arrive in two distinct stages, the initial period of rain starting with the arrival of altostratus cloud and the second period starting when the surface front is much closer.)

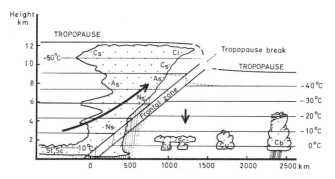

Fig. 8.10. Vertical cross-section through idealized warm front. (Note that the vertical scale is exaggerated by a factor of over 100.)

Air motions *relative to* our idealized warm front are indicated by arrows in fig. 8.10, although the presence of such motions is in part demonstrated by the accompanying cloud sequence—the main frontal cloud belt being associated with the upward motion in the warm air, and the damping down and spreading out of the cumulus beneath the frontal zone resulting from the subsidence and greater stability of the cold air in that region. Note that these relative motions imply also a distinct convergence† of air towards the front: for this reason the normal speed of the front, V_F, is less than that implied by geostrophic considerations, $(V_G)_n$, fig. 8.5. For a warm front $V_F \sim 70\%$ of $(V_G)_n$.

The temperature structure of our idealized front is shown by a series of solid isotherms in fig. 8.10. In reality, about half of the total air mass contrast is concentrated in a region usually recognizable from upper-air observations as a frontal zone, while the total contrast occurs over a region perhaps ten times as great in extent, as indicated by the broken isotherm for $-40°C$. Irregularities of ascent, descent, evaporation and precipitation occur, especially near the edges of frontal cloud,

† This air is removed at upper levels by a (thermal) wind component (not shown) which blows parallel to the front and out of the plane of the paper.

and hinder the clear recognition of the frontal zone.   In particular, the cloud-free air, which marks the boundary of the jet stream on its cold air side, is often remarkably dry, suggesting that it may have descended from the stratosphere through the 'tropopause break' shown on the diagram.   This cloud-free air implies that the slope of the frontal cloud is greater than that of the front, as shown in fig. 8.10.

The cloud structure of an individual front depends on the moisture content and stability of the interacting air masses and on their modification by vertical motion near the front.   If, for example, the warm air mass is potentially unstable then this instability may be released in the region of upward motion above the frontal zone, with the result that cumuliform rather than stratiform clouds, perhaps even thunder, are associated with the front in question: this is very rare in the British Isles.

It would be wrong to conclude from fig. 8.9 that the vertical motion associated with a warm front can always be viewed as the sliding of warm 'light' air over cold 'dense' air.   The sense and distribution of vertical motion in the vicinity of a front depend on the detailed temperature distribution in the frontal region, on the shape of the isobaric trough containing the front and on the slope of the frontal surface.   It is theoretically possible to have fronts with cloud systems and precipitation associated with upward motion in both air masses, or even in the cold air mass alone.   In practice, however, the cold air mass usually lacks sufficient water content to maintain an extensive cloud system, while the limited depth of such a system severely restricts its ability to produce precipitation: a frontal surface is usually an effective barrier to vertical motion.

The more active of two main types of cold front, usually called the ana cold front, is very similar in structure to the warm front but has a frontal slope two or three times greater.   Thus the mirror image of fig. 8.10 may be taken to represent an idealized ana cold front, provided that the frontal slope of 1 : 150 is replaced by one of 1 : 60 or so; this means that there is, typically, a rain belt some 200 km wide behind the front.

The much less active kata cold front has quite different characteristics. The air descends not only within the cold air mass but also in the warm air close to the frontal surface.   Forced uplift of the warm air ahead of the surface front causes clouds to form, but they are seldom more than a few kilometres in depth and cause only light precipitation.   The cloud breaks when the surface front passes—sometimes even earlier.

8.4. *Frontal depressions*

The development of most mid-latitude depressions occurs along the front separating polar air from tropical air—the polar front; while depressions originating at high latitudes usually do so at some point on the arctic front which separates polar air (frequently rPm) from arctic air.

8.4.1. *The life cycle of a frontal depression*

Most depressions which form on the polar and arctic fronts pass through a life cycle much as indicated by the sequence (*a*) to (*e*) on fig. 8.11. Frontal lows in various recognizable stages of development are present in figs. 8.1, 8.2 and 8.12. The lifetime of an individual frontal low is usually from three to perhaps five days, during which it travels several thousand kilometres (perhaps even 10 000 km): however, stage (*c*) is often attained within 24 hours.

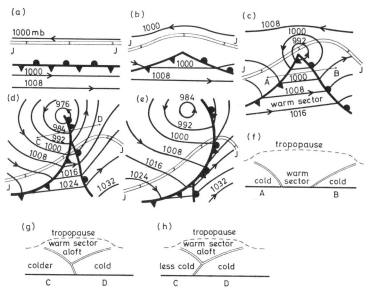

Fig. 8.11. The frontal depression. (a) to (e); life cycle stages. (f) to (h); vertical cross-sections: (f) of the ‘open’ warm sector in (c): (g) and (h) of the occluded front in (d). (g), cold occlusion. (h), warm occlusion. JJ: jet stream axis.

In explanation of fig. 8.11, consider (*a*) to represent an essentially straight, quasi-stationary part of the polar front, orientated in a roughly east–west direction. The cyclonic wind shear implied by easterly winds to the north of this front and westerly winds to its south enables the front to remain in an overall equilibrium state, provided that no ‘outside’ influence causes a local distortion of the front in the sense shown in (*b*). Such a distortion, called a frontal wave, always travels in the same direction and with much the same speed as the warm air on its southern flank. The leading edge of the wave is a warm front and its trailing edge a cold front.

Most frontal waves are unstable in the sense that they tend to increase in amplitude; at the same time pressure falls near the tip of the wave and closed isobars may soon be drawn, as in (c). The region enclosed by the warm and cold fronts is known as a warm sector and the disturbance at this stage as a warm-sector depression.

A large majority of the lows which reach the warm-sector stage deepen further and become more vigorous. The cold front travels faster than the warm front, overtaking it first of all near the centre of the low and then progressively farther from the centre, as indicated in (d) and (e), fig. 8.11. This process is known as occlusion and clearly is associated with the raising aloft of the warm-sector air.

The air in the cold sector of the depression is initially homogeneous. However, the processes that modify the air in advance of the warm front differ somewhat from those that modify the air behind the cold front; a new front is therefore formed when the occlusion process brings these air masses together. The front is itself called an occlusion and may be of the warm or of the cold type, as illustrated in figs. 8.11 g and h. One or the other of these is appropriate to a vertical cross-section along the line CD in fig. 8.11 d. The tendency is for occlusions, which move overland from the Atlantic, to be of the warm type in winter and of the cold type in summer. The warm sector cross-section shown in fig. 8.11 f refers to the line AB in fig. 8.11 c.

Before a frontal depression occludes, its velocity is about that of the geostrophic wind in its warm sector. After occlusion has well started, the centre of the low usually (but not always) turns off to the left of its initial track, and also begins to slow down. This is the stage of lowest central pressure and strongest winds.

The final stage in the life of a frontal depression is represented by fig. 8.11 e. The process of occlusion is virtually complete; the occlusion itself has become detached from the centre of the low, which has no longer any preferred direction of motion. The low fills up rapidly if a more vigorous depression moves into the area. On other occasions— especially those that mark the end of a sequence—it persists almost unchanged for days, fresh intrusions of polar air appearing to sustain it against the dissipating effects of surface friction. A thermal low almost overlies the surface low so that the circulation is deep and any depressions which may approach from the west are guided round its southern flank.

The life cycle of the frontal depression represents a conversion of potential energy into kinetic energy. Initially, solar energy is converted into atmospheric heat energy. The distribution of this is uneven: in particular, air masses of different densities often lie side by side along a front in middle latitudes, representing a concentration of available potential energy there. As the frontal depression develops, potential energy is converted into kinetic energy, as represented by strengthening winds over a large area. Finally, cold air of high density completely underlies warm air of low density and there is no longer any available

136

potential energy.   Surface friction acts as a brake on the winds through-out and finally brings the circulation to a halt†.

8.4.2. *Cold front waves; depression families*

New wave depressions are frequently observed to form on the trailing cold front of an occluded low.   In many cases, these move rapidly along the front without significant amplification and are absorbed by the parent low, or are progressively damped out and fail to reach the parent low; such cold front waves are said to be stable.   Their main effect, other than causing a local intensification of the precipitation associated with the cold front, is to retard the progress of the front itself and thus delay the clearance usually associated with its passage: this last is perhaps the most important consideration from the point of view of weather forecasting.

By no means all cold front waves are stable.   In particularly favour-able conditions, some develop so rapidly that they completely absorb their own parent depression within a matter of 24 hours.   More usually, however, development is much less rapid than this.

Figure 8.12, appropriate to 1200 G.M.T. on 15 October 1967, demonstrates a not uncommon synoptic situation in which the polar front stretches across perhaps a quarter of the globe.   The sequence of depressions marked A, B, C, D, E, and F is known as a depression family. Each successive member of a family develops from a wave on the cold front of its immediate predecessor.   Also, it usually moves on a track a little to the south of that followed by its predecessor so that the eventual tendency is for the polar front to encroach on the sub-tropical high pressure belt.

8.4.3. *Warm front waves*

Much less commonly, wave depressions form on warm fronts.   The initial indication on the surface chart is the slackening of the gradient across a particular region of an extended warm front.   They hardly ever deepen beyond one or two isobars, invariably move along the general direction of the warm front, away from the parent low‡, and fill up as they penetrate the higher pressure which is located there.   Their development and passage cause much more widespread cloud and precipitation than would otherwise occur on a warm front at so great a distance from the parent low.

† Friction would bring the atmosphere almost to rest in a few days if the kinetic energy were not renewed by solar energy.

‡ A frontal depression, whether primary or secondary, moves with a velocity approximating to that of the thermal wind component derived from the gradient of the 1000 to 500 mb thickness above its centre.   This is known as 'thermal steering'.

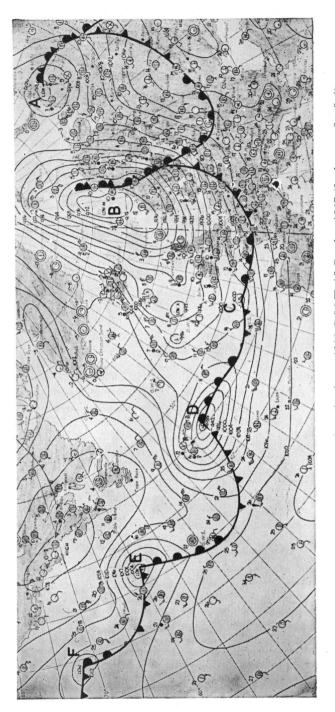

Fig. 8.12. Family of depressions on the polar front; 1200 G.M.T. 15 October 1967 (for key see fig. 8.1).

138

8.4.4. *Secondaries at points of occlusion*

A secondary depression sometimes forms at the ' triple point ' on the surface chart at which the occlusion and the warm and cold fronts meet, the developments being quite different for the warm and cold types of occlusion.

The warm occlusion case strongly resembles that of the warm front wave. On the surface chart a warm occlusion is a continuation of a warm front, the usual alignment being from about north-west to south-east. If, to the east of the fronts, the north-westerly thermal wind component is unusually strong at the stage when the primary depression becomes slow-moving, then an occlusion secondary may form. On the surface chart it is seen initially as an opening-out of the isobars near the point of occlusion, followed by a concentration of pressure fall in this region and a fairly rapid movement of the secondary, which seldom deepens much, along the general direction of the front and away from the parent low.

In the case of the cold occlusion the alignment of the cold front and occlusion on the surface chart is usually from about north-east to south-west. If the upper winds to the west of the cold front are (near) jet stream speed and if the streamlines fan out near the point of occlusion when the primary depression is becoming slow-moving, then a secondary may well form at this point. Its usual movement after formation is towards the east at a slow or moderate speed and it may deepen considerably.

8.5. *Non-frontal depressions*

There are several types of depression which do not originate as frontal waves.

8.5.1. *Heat lows*

Shallow depressions often develop over islands and peninsulas during daytime, especially during clear summer weather. Their origin lies in the widely different thermal inertias of land and sea, and may be explained as follows.

Consider fig. 8.13 *a* to represent a large island in the early morning, in cloudless conditions. Assume that temperature and pressure are uniform at all levels over the entire region, so that the chosen pressure levels, 1000 and 850 mb, are initially horizontal. During the day the land surface becomes much warmer than the surface of the surrounding sea, and this warmth is communicated to the air above the island. The result is indicated in figure 8.13 *b*, in which it is assumed that the atmosphere above 850 mb is unaffected by diurnal temperature variations, so that the 850 mb pressure level is the same as in fig. 8.13 *a*. However, the thickness of the 1000–850 mb layer is increased by the warming of the island so that the 1000 mb pressure level dips down towards the surface.

Figure 8.13 c shows likely mean-sea-level isobars for the region in mid-afternoon.

The associated circulation pattern is indicated in figure 8.13 b, the inward flow of cool air at low levels being the familiar sea-breeze. Initially this pattern is radial in the sense implied by the solid wind arrows in figure 8.13 c, but the converging air soon acquires a cyclonic spin—see § 8.7.2. (Likewise the outflowing air acquires an anti-cyclonic spin so that a 'heat high' exists in the upper part of the layer.)

Associated weather very often remains sunny. However, if the air is relatively unstable the low-level convergence may provide the trigger required for afternoon thunderstorms.

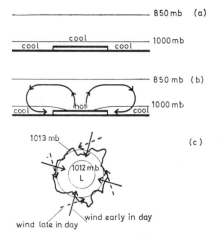

Fig. 8.13.   The heat low.   Sequence (a) and (b) shows the formation of a heat low over an island, (c).

8.5.2. Polar lows

When polar or arctic air flows out over the ocean, and in consequence is heated intensely from below, small depressions known as polar lows are frequently observed to form, especially in winter when the sea-minus-air temperature difference is greatest. As this temperature difference is maintained day and night, a polar low is not a diurnal feature like a heat low; also it does not remain anchored over a particular region, as a heat low does over land, but is carried along by the Pm or Am airstream in which it is embedded.

The most favourable region, synoptically, for the development of a polar low is to the west of a large occluding low. Three polar lows appear in just such a position in fig. 8.12 (to the south of Iceland).

Cumulonimbus clouds and heavy showers of rain, hail or snow are associated with polar lows. In winter these showers may merge to give prolonged outbreaks of snow. If the airstream carrying the polar low is fast-moving, fierce squalls may accompany individual showers. (A polar low associated with the Am airstream in fig. 8.2 was responsible for a British Isles record sea-level gust of 118 kt in the Orkney Islands on 7 February 1969.)

Although polar lows are clearly caused by strong surface heating of an unstable environment, the precise mechanism which leads to their formation is not yet fully understood.

8.5.3. *Orographic lows*

The effect of a transverse mountain barrier on a broad airstream is to distort it in the manner shown in fig. 8.14 *a*: i.e. there is a ridge upwind of the range and a trough in its lee. Behind a high mountain range, however, a closed centre of low pressure, or orographic low†, may exist, as shown in fig. 8.14 *b*. Such a centre is almost stationary, its precise position varying with the direction of the incident airstream.

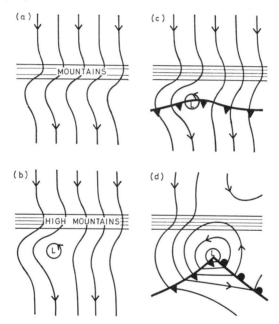

Fig. 8.14. The orographic low. (a) Formation of an orographic trough. (b) Formation of a closed centre of low pressure. (c) Distortion of a passing cold front, with the generation of a frontal wave depression (d).

Owing to the downward component of flow in the lee of a mountain barrier an orographic low seldom produces any active weather phenomena. However, it may well be activated during the passage of a front (especially a cold front). When this occurs, the front is usually retarded by the presence of the orographic low and is distorted by its circulation

† Also called a lee depression.

(fig. 8.14 c).   At this stage it is possible for an orographic low to deepen and assume the characteristics of a frontal wave depression, as indicated in fig. 8.14 d.   Such a sequence of events is frequently observed on the passage of a cold front southwards over the Alps in winter, the associated orographic low being in the region of the Gulf of Lyons.

8.5.4. *Tropical cyclones*

Over the tropical oceans, cumulonimbus clouds are associated mainly with the ITCZ in which the winds from the two hemispheres meet (see fig. 10.1) but they do occur elsewhere, especially in the more western parts of the oceans and well away from the sub-tropical anticyclones. Local convergence of surface winds occurs near a group of cumulonimbus clouds and may develop an organized cyclonic circulation.   In conditions which are not well understood, but are usually associated with a wave or trough in the tropical easterly flow, this circulation grows both in strength and in horizontal extent and, although development may cease at any stage, intense cyclonic storms may result.   These are called hurricanes in the western Atlantic, and typhoons in the western Pacific, and have rather special features.   They are much more concentrated than higher-latitude depressions, their associated winds being among the most destructive of natural phenomena: by definition these attain at least hurricane force (force 12 on the Beaufort scale of force, i.e. 64–71 kt) but may be nearly twice as strong.

Hurricanes form within homogeneous air and so lack the source of potential energy which allows frontal depressions of higher latitudes to develop.   However, the energy supplied by the warm tropical oceans compensates for this lack†.   Fully-developed hurricanes are almost confined to later summer and autumn when the sea surface temperatures are at their highest.   They do not form nearer the equator than about 5° because of the virtual absence of the Coriolis force there.

As the air converges towards the low pressure centre it is made very moist by rapid evaporation into it from the warm ocean.   The convergence causes the air to spin faster and it spirals inwards and upwards at an increasing rate (see § 8.7.2), causing torrential rain and thunderstorms.   The latent heat released by condensation of the water vapour warms the air and keeps it unstable.   The clouds are mainly cumulonimbus and are arranged in spirals round the centre (as is very evident from satellite pictures).   Near the tropopause the air diverges outwards from the centre.

The most individual feature of the hurricane is its central 'eye', typically some 40 km across.   Most of the air which spirals inwards does not reach this central region of lowest pressure which is characterized by an absence of rain, light winds, descending motion, and warm air most of

† The rapid decline of hurricanes after passage inland shows their dependence on latent heat supply.

which has subsided from higher levels.   In contrast, the strongest winds and heaviest rain occur just outside this central eye.

Hurricanes usually move slowly.   They have velocity components towards the west and away from the equator at first, and often recurve polewards and towards the east when they reach 20°–30° latitude. Some of the most intense depressions to affect the British Isles in autumn originate as hurricanes; in such a case the area covered by the cyclonic circulation increases gradually as the storm moves eastwards across the Atlantic.

8.5.5. *Tornadoes*

The ultimate stage of concentration of both cyclonic spin and energy of atmospheric motion is reached in the tornado†.   When a cumulonimbus cloud is in the process of rapid development the air which converges towards its base spirals upwards about a core of reduced pressure.   If the air is particularly humid, water vapour may condense in the core and a ' funnel cloud ' then projects down from the main cloud base. Fresh energy is released by this condensation and the cloud may then extend to ground level to form the tornado: the cloud column is usually twisted—hence the name ' twister ' sometimes given to the phenomenon.

Tornadoes average about 100 metres across.   Their destructive power is caused both by their tremendous winds, which may exceed 200 kt, and by the great radial gradient of pressure which causes explosive destruction of buildings etc. in their path.   Their duration is very variable, about 10 to 15 min on average.   They occur in many parts of the world including, very rarely, the British Isles.   They are perhaps most common in the central U.S.A. where their maximum reported frequency is about 1 per year per 80 km square.

8.6. *Anticyclones*

Continuity requires that, on every scale of atmospheric motion, each type of circulation has its counterpart.   For the depression or cyclone this is the anticyclone.   The three-dimensional aspects of the counteracting motions involved are examined in § 8.7, where it is also shown that the systems are formed by dynamical processes of essentially the same type, each the necessary complement of the other.   It is therefore not surprising that the general structure and associated weather of the anticyclone, which is the concern of this section, are the complete antithesis of those of the depression.

8.6.1. *General characteristics*

In an anticyclone there is descending motion through much of the troposphere.   This motion compresses the air and causes it to warm; since no water vapour is added, the relative humidity of the air is

---

† The tornado occurs over the land, the corresponding (though less violent) feature over the sea being the waterspout.

decreased and any clouds which were initially present are dissolved. The air mass properties are almost uniform over a region which is often very large (up to about 10 000 km in horizontal extent in some cases). The central regions of the anticyclone have very light winds, there being a theoretical maximum wind speed (in anticyclonic curvature) which increases with distance from the centre (§6.3). On the surface chart the anticyclone is usually stationary or slow-moving.

Despite the general dryness and absence of cloud in much of the troposphere, anticyclonic weather is not uniformly fine. This is because the rate of general descent of air decreases in the lower part of the troposphere; the air within the lowest kilometre or so has not descended and is mixed vertically by turbulence. If this air is moist a layer of stratus or stratocumulus may be formed. Especially in winter there is a tendency for the cloud to be ' all or nothing '†. Once formed in this season it does not disperse easily because the cloud top cools by outward radiation and so tends to thicken slowly upwards, thereby forming an increasingly sharp inversion with the dry subsided air immediately above (a temperature inversion acts as a barrier to vertical motion).

In winter spells of such ' anticyclonic gloom ' the pollution products of industrial regions are prevented from escaping upwards. If the anti-cyclonic inversion is unusually low and the anticyclonic spell is protracted, the concentration of the products at the ground may reach an un-comfortably, even dangerously, high level. These are occurrences of so-called smog. Severe winter frosts occur in anticyclones in which the surface air is relatively dry and there is no widespread fog.

Anticyclonic spells in summer are usually fine and provide the conditions which yield the highest maximum temperatures. There is continuous warming by subsidence in the development stage of the anticyclone. The inversion which is formed between the subsided air and the air in the unsubsided layer at the earth's surface may be low enough to merge with an overnight radiation inversion. The solar energy subsequently absorbed during the day under cloudless conditions is thus confined to a shallow layer, permitting the surface temperature to attain exceptionally high values.

Stratocumulus forms at times in summer anticyclones but is less persistent and has a higher base than in winter. However, conditions sometimes favour the formation of coastal fog by advection if the track over the sea is sufficiently long.

8.6.2. *Cold and warm anticyclones*

The excess pressure of an anticyclone arises from the presence of cold dense air, relative to the surroundings, in a confined layer. On the basis of the level at which this cold air is found, anticyclones are classified either as cold or as warm.

† In middle latitudes a winter anticyclone over the sea is usually cloudy.

The former is cold at low levels, as in the winter anticyclones formed over Siberia and Canada by protracted radiational cooling of the lower layers. Since pressure decreases with height more rapidly in cold air than in warm, an anticyclone of this type decreases in intensity with height. Thus the cold anticyclone may only be evident below a height of 2 km; above this level the circulation may in fact be reversed, though more commonly the surface anticyclone is overlain by strong upper westerlies.

At the other extreme is the warm anticyclone, characterized by relatively warm air in the low and middle troposphere (the circulation therefore increasing with height in this region). The excess surface pressure arises in this case from a high tropopause level: the air in the higher troposphere is therefore relatively cold (as it is near the equator relative to higher latitudes—see fig. 2.2) and more than cancels the effect of the warm air at lower levels. The sub-tropical anticyclones (see §10.2.1.) are of this warm type.

Not all anticyclones are definitely of one type or the other. The highest mean-sea-level pressures of all (in excess of 1070 mb) occur when, in the continental anticyclone, cold air at low levels happens, exceptionally, to be combined with a higher-than-average tropopause. Again, the mobile anticyclones that sometimes form in the rear of a cold front of a mid-latitude depression are initially cold but are gradually transformed into the warm type as the low-level air warms and the tropopause rises. These anticyclones usually move towards, and merge with, the sub-tropical anticyclones after an initial eastward motion.

8.7. *Synoptic development*

The changes in the synoptic situation from day to day, and the sequence of local weather, involve both the movement of pressure systems and their development. Although these are not independent—movement is controlled to some extent by development and vice versa—it is convenient for some purposes to separate them. In this section we look for an explanation of why cyclones and anticyclones develop, starting first with some general principles which involve two related concepts of fundamental importance in development, namely divergence and vorticity.

8.7.1. *Convergence, divergence and vertical motion*

(1) Meaning and relationships

Figures 8.15 *a* and *b* illustrate various patterns of horizontal convergence and divergence of flow. The first of these—direct convergence towards a point or divergence from it—is more extreme than occurs in nature but the others are realistic. Horizontal convergence or divergence implies that the separation, in the horizontal plane, of individual particles of air is changing with time; the magnitude of convergence (or divergence) is the fractional decrease (or increase) of area enclosed by a given chain of particles in unit time. To fix ideas: if, in the second diagram of fig. 8.15 a, the indicated radius is 100 km and the winds have a speed of 10 m s⁻¹ and cross the circle at an angle of 30° then the inward velocity is 5 m s⁻¹ and the convergence is calculated to be very nearly 10^{-4} s⁻¹ $= 3.6 \times 10^{-1}$ hr⁻¹. Similar calculations may be made for the other types of flow, with the help of appropriate formulae to be mentioned later.

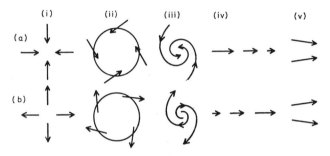

Fig. 8.15. Patterns of (a) convergent and (b) divergent horizontal motion: (i) motion to or from a point; (ii) cross-isobar flow; (iii) spiralling motion; (iv) down-stream change of speed; (v) confluence and diffluence.

It is observed that when the streamlines in the upper atmosphere are confluent (suggesting convergence) the wind speed usually increases downwind (suggesting divergence); and conversely, when the streamlines are diffluent the speed decreases downwind†. Thus the divergence or convergence at a particular point in the upper atmosphere is often a small difference between two larger but counteracting effects: in larger-scale motion it seldom exceeds 10^{-5} s⁻¹.

If an incompressible fluid, such as water, is squeezed horizontally it expands vertically by the amount required to keep the total volume and mass constant. With a compressible medium such as air there is the further possibility that horizontal convergence may cause the density to increase (or that divergence may cause the density to decrease). Observations show that changes of air density of this kind are small and that vertical squeezing of a layer of air is, to a large extent, compensated by its horizontal expansion (as was mentioned for a subsiding layer in §5.4.1), and vertical stretching by horizontal compression. For an atmospheric layer, vertical stretching implies that Δp, the pressure difference between the top and bottom of the layer, increases and vertical shrinking implies that Δp decreases.

† If the streamlines were isobars and the flow geostrophic then the speed would increase in direct proportion to the decrease in the separation of the isobars. These are not, however, conditions when strict geostrophic control would hold.

146

The main interest lies in the horizontal divergence of velocity (div_H **V**), usually therefore referred to simply as the divergence. Where vertical divergence is intended it is referred to as such, and the three-dimensional divergence (which is almost zero) may be referred to as the total divergence (div **V**). Positive divergence implies negative convergence and vice versa, i.e. div_H **V** = $-\text{con}_\text{H}$ **V**: equations are always expressed in terms of divergence.

(2) Vertical distributions

The foregoing principles may be applied to the (apparent) anomaly that charts such as figs. 8.1 and 8.2 show a systematic cross-isobar flow of air from anticyclones to depressions. Since these transfers by themselves would quickly destroy the systems how do the latter, in fact, survive and even intensify?

The air which converges towards depressions at low levels cannot accumulate there without causing a continuous increase in density, and it is therefore forced to ascend. With continuous supply from below, this air cannot remain within the same vertical column (otherwise pressure would rise continuously) and it therefore diverges horizontally at higher levels; the evidence is that this occurs in the high troposphere. The resulting exchange of air between adjacent depressions and anticyclones is shown schematically in fig. 8.16; the descending motion in

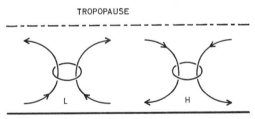

Fig. 8.16. Schematic exchange of air in the troposphere between adjacent depressions and anticyclones.

anticyclones and ascending motion in depressions explain their general weather features. The pattern of vertical motion and divergence reverses again in the stratosphere, perhaps more than once, but the effect on the lower part of the atmosphere is small because of the low density of the air in these upper regions.

(3) Surface pressure changes

For both types of system the surface pressure falls if the divergence of mass exceeds the convergence of mass, each summed through the whole atmospheric column†. This happens during about the first half of the

† Divergence of mass is not the same as divergence of velocity because the former includes a weighting factor for the densities of particles involved in the motion. The distributions of the two forms of divergence are the same if there are no horizontal gradients of density; in general, they are nearly the same.

life cycle of a depression, and the surface convergence, upward motion and high-level divergence are then all strong. In the later stages, when the central pressure is rising, the upper-level divergence is either absent or weakens to the extent that it does not compensate the surface frictional convergence (which is also then weakening because of decreasing winds). The mechanism of these pressure changes in earlier stages of depressions is made more complicated by the strong thermal gradients that then occur. The horizontal uniformity of temperature near anticyclones simplifies matters. Upper-level convergence and subsidence of air are strong when the anticyclone is intensifying and die out in later stages.

(4) Some formulae

In physical terms, div $\mathbf{V}$ and $\text{div}_H \mathbf{V}$ are the fractional increases of volume (V') and horizontal area (A), respectively, enclosed by chains of particles, in unit time, i.e.

$$\text{div } \mathbf{V} = \frac{1}{V'}\frac{dV'}{dt} \tag{8.8}$$

and

$$\text{div}_H \mathbf{V} = \frac{1}{A}\frac{dA}{dt}. \tag{8.9}$$

Where u, v, w are the velocity components in the x (west-to-east), y (south-to-north) and z (vertically upwards) directions, the total divergence is

$$\text{div } \mathbf{V} = \frac{\partial u}{\partial x} + \frac{\partial v}{\partial y} + \frac{\partial w}{\partial z} = \text{div}_H \mathbf{V} + \frac{\partial w}{\partial z}. \tag{8.10}$$

The horizontal divergence ($\text{div}_H \mathbf{V}$) is related approximately to the vertical divergence ($\partial w/\partial z$) by the equation

$$\text{div}_H \mathbf{V} \equiv \frac{\partial u}{\partial x} + \frac{\partial v}{\partial y} = -\frac{\partial w}{\partial z}. \tag{8.11}$$

$\text{Div}_H \mathbf{V}$ may be calculated, at any point, from the distribution of u, v in the region concerned, finite differences being used. For the example previously quoted (inflow of 5 m s^{-1} into a circular area of radius 100 km) both u and v have a difference of 10 m s^{-1} in x and y distances of 200 km and thus

$$\text{div}_H \mathbf{V} = \frac{\Delta u}{\Delta x} + \frac{\Delta v}{\Delta y} = -\frac{10}{2 \times 10^5} - \frac{10}{2 \times 10^5} = -10^{-4}\ \text{s}^{-1}.$$

Calculations of divergence in actual cases show that the numerical value is very sensitive to small changes in the distribution of wind speed and direction and that the latter is not accurately enough known, in general, to make such calculations by direct means worth while. In view of the importance of divergence in development and in its relation to vertical velocity (equation (8.11)), it is fortunate that it may be assessed indirectly from the spin (vorticity) of the air, as will now be shown.

8.7.2. *Convergence a::d vorticity*

(1) Experiment

The importance of the convergence and divergence of fluid particles in relation to changes in their vorticity is well illustrated by the familiar bath-plug vortex—a phenomenon that may also be observed in a laboratory or kitchen sink.   If the water is stirred, clockwise or counter-clockwise, and the plug is then removed a strong vortex with the same sense of rotation as that imparted forms round the hole.   If the water is motionless before the plug is removed no vortex is formed.† These observations show that the convergence of the water particles towards the hole concentrates any existing spin of the particles but does not create spin if none exists before convergence.

These observations illustrate the law of conservation of angular momentum (see also §10.3).   Before the plug is removed, individual particles have an angular momentum (per unit mass) given by the product of their angular velocity and their radial distance from the hole.   As they converge towards the hole (so decreasing the radial distance) their angular velocity increases, in the absence of friction or other force acting round the axis, by the amount required to keep this product constant; the actual increase in angular velocity is less than this because of friction, as is true of low-level flow in the atmosphere.   A further good illustration of the principle is the way in which a skater controls his rate of spin by adjusting the positions of his feet, legs and arms relative to his body.

(2) Relative and absolute vorticity

The vorticity of the solid Earth about the local vertical in latitude ϕ is twice its angular velocity, $\Omega \sin \phi$: it is therefore $2\Omega \sin \phi$ or f (the Coriolis parameter) and is cyclonic in sense (see Chapter 6).   The air shares this vorticity and has, in general, an additional spin relative to the earth, denoted ζ (Greek zeta), which may be cyclonic or anticyclonic. Thus the air's vertical component of relative vorticity is ζ and of absolute vorticity is $(\zeta+f)$, ζ having the same sign as f when it is cyclonic. Because large-scale motion is nearly horizontal it is only this component, which represents a spin in the horizontal plane, that is of concern.   It will be referred to simply as the vorticity.

Simple horizontal curvature and simple horizontal shear of motion each involves cyclonic or anticyclonic rotation of air particles. ζ arises, in general, from a combination of curvature and shear effects: in some

† A systematic vortex may be formed because of asymmetry of shape (exerting an effective force round the hole) but this may be reversed by pre-stirring in the opposite sense.   (Under the most rigorous experimental conditions the Coriolis force has been shown to have an effect.)

instances these components reinforce each other, while in others they are in opposition. Various cases are illustrated in fig. 8.17.

ζ is a field quantity which is determined, at any point, by the distribution of velocity about that point. There are two alternative formulae for ζ, the second of which (equation (8.13)) is applied in

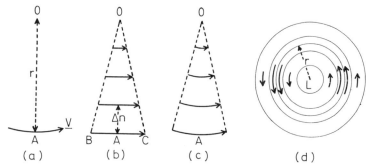

Fig. 8.17. Curvature and shear contributions to vorticity. (a) Curvature only (cyclonic): e.g. for $V = 20$ kt and $r = 200$ n. miles, $\zeta_A = V/r = +0.1$ hr^{-1}. (b) Shear only. Winds are stronger on right of current than on left and particles rotate from line OB to line OC i.e. in cyclonic sense (northern hemisphere): e.g. if shear is uniform and $\Delta V = 5$ kt over distance $\Delta n = 50$ n. miles, $\zeta_A = -\Delta V/\Delta n = +0.1$ hr^{-1} (Δn measured to left of wind). (c) Curvature and shear. The effects combine and $\zeta = +0.2$ hr^{-1}. V/r is constant in this particular case and the air rotates as a solid body with a vorticity twice its angular velocity. (d) Distribution of contours or streamlines in which curvature is cyclonic everywhere: shear is cyclonic out to distance r from the centre, then anticyclonic at greater distances. (Middle latitude depressions have roughly this form of velocity distribution.)

In all cases the sign of the vorticity is reversed if the winds are reversed.

fig. 8.17 in the finite difference form which is applied in practice. The formulae, which give positive values for cyclonic vorticity, are

$$\zeta = \frac{\partial v}{\partial x} - \frac{\partial u}{\partial y} \qquad (8.12)$$

and

$$\zeta = \frac{v}{r} - \frac{\partial v}{\partial n}, \qquad (8.13)$$

where u, v are the velocity components in the x, y directions and, in equation (8.13) (which expresses ζ as the sum of curvature and shear), r is the radius of curvature of flow and distance n is measured normal to and to the left of the wind.

If, therefore, the field of motion is known, the meteorologist can calculate the corresponding distribution of relative vorticity ζ and may represent it by isopleths, i.e. by lines joining places of equal ζ. In general, this resembles the isobaric or height contour fields, modified however by the effects of wind shear which, in some circumstances, are very important. If the corresponding value of f (a function of latitude only) is added algebraically to each value of ζ and contours are then drawn to these revised figures, the field of absolute vorticity is represented. Because cyclonic ζ has the same sign as f and also has numerically higher values than anticyclonic ζ, positive values of $(\zeta+f)$ are almost universal. However, small negative values of $(\zeta+f)$ do occur in regions of strong anticyclonic shear of wind in the upper atmosphere; also both ζ and f tend towards zero at the equator.

(3) Formation of cyclones and anticyclones

If, now, the results of the observations of the relation between convergence and vorticity are applied to the atmosphere, the significance of absolute vorticity may be appreciated. First, the reason that $(\zeta+f)=0$ at the equator is that, since $f=0$ there, equatorial air possesses no initial spin to be augmented by convergence. So pressure systems are not found near the equator.

The general relationship between divergence and vorticity may be expressed in the formula

$$\frac{1}{\zeta+f}\frac{\mathrm{d}}{\mathrm{d}t}(\zeta+f) = -\operatorname{div}_{\mathrm{H}}\mathbf{V}: \qquad (8.14)$$

i.e. the horizontal convergence of the wind velocity equals the fractional increase of its absolute vorticity with time; alternatively, horizontal divergence produces a decrease of absolute vorticity with time.

Near the Earth's surface $(\zeta+f)$ is positive so that any convergence there creates cyclonic vorticity (neglecting change of f). However, as was explained, this cannot proceed unless there is a means of removing the air laterally at higher levels. The key to the development of cyclones and anticyclones, at least in middle and higher latitudes, lies in the increase of westerly winds with height, culminating in the polar front jet stream in the high troposphere.

An association of jet streams with depressions was shown in fig. 8.11. The air moves rapidly through the jet stream pattern, speeding up as it enters and slowing down as it leaves. These changes of speed, and also the shear that exists across the stream, mean that the motion is unbalanced and that the high-level divergence and convergence required for the development of depressions and anticyclones, respectively, occur there.

Curvature of flow, greater or smaller in degree, is a universal feature of jet streams. Because of the effect of curvature on wind speed (see § 6.3) the air moves at less than the geostrophic speed through a trough and at more than the geostrophic speed through a ridge. So the air tends to pile up on the upwind side of the trough (favouring a surface high below) and tends to diverge on the downwind side of a trough (favouring a surface depression).

151

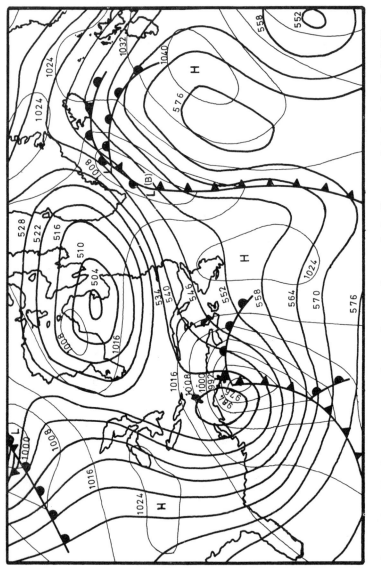

Fig. 8.18. 500 mb contours superimposed on surface analysis (fronts and isobars) for 00 G.M.T. on 10 February 1969. Contours are numbered in tens of metres.

152

Since the contour patterns at 500 mb are usually not very different from those at jet-stream level, about 200 mb, it is of interest to see how the 500 mb contours (available in the upper-air section of the Daily Weather Report) are related to surface pressure.   In fig. 8.18 note that the depression which lies off the east coast of the U.S.A.† is downwind from a trough in the contours; also that an anticyclone lies upwind from this trough.   Similar features may be seen in other parts of the chart. The contour chart is 12 hours later than the surface chart shown in fig. 8.2: the Atlantic depression (Low B) on the latter chart has meantime become much less deep in an ' unfavourable ' contour pattern and it is the low which lies west of Iceland, upwind from the 500 mb contour ridge, which subsequently deepens. These various features are in agreement with the general rule deduced above.

(4) Potential vorticity

We have seen earlier that the horizontal divergence of a column of air implies that it shrinks vertically and also that it decreases its absolute vorticity; and, conversely, that horizontal convergence is associated with vertical stretching of a column and an increase of its absolute vorticity. We may perhaps infer that, as a column of air moves along, it shrinks vertically just in the proportion that it decreases its absolute spin, or it expands vertically in proportion to the increase in its absolute spin; i.e. where Δp is the pressure difference between the top and bottom of the column,

$$\frac{\zeta+f}{\Delta p} = \text{constant (following the motion)}.$$

This is an approximate relationship which is easily derived from equations (8.9) and (8.14) as follows:

$$\text{div}_\text{H} \mathbf{V} = \frac{1}{A} \frac{\mathrm{d}A}{\mathrm{d}t} = -\frac{1}{\zeta+f} \frac{\mathrm{d}(\zeta+f)}{\mathrm{d}t};$$

$$\therefore \quad A \frac{\mathrm{d}}{\mathrm{d}t}(\zeta+f) + (\zeta+f)\frac{\mathrm{d}A}{\mathrm{d}t} = 0.$$

Thus, integrating,

$$A(\zeta+f) = \text{constant}.$$

But for an atmospheric column $A\Delta p \sim \text{constant}$.

$$\therefore \quad \frac{\zeta+f}{\Delta p} = \text{constant}.$$

The value of $(\zeta+f)$ that corresponds to a standard value of Δp(50 mb, say) labels a column of air just as the potential temperature (θ) labels an element of air in dry adiabatic motion (Chapter 3).   This property has been used in modern numerical weather forecasting.   It also offers a

† This depression was the cause of a particularly severe snow-storm in east New York State.

dynamical explanation of the orographic trough or low mentioned earlier. Consider, for simplicity, the passage of a uniform westerly air stream ($\zeta = 0$) across a north-to-south mountain range (fig. 8.19). The air speeds up on ascent and slows down on descent. Each column has therefore an increasing horizontal spread and a decreasing pressure depth Δp as it ascends, and a decreasing horizontal spread and increasing Δp as it descends. It follows from the above theorem that $(\zeta + f)$ decreases on ascent, i.e. ζ, previously zero, becomes anticyclonic and the air curves southwards. Beyond the summit, Δp and $(\zeta + f)$ increase. Since f is decreasing at this stage because of the southward motion of the air induced earlier, ζ first becomes less anticyclonic, then cyclonic, so that the air turns again to the north and a lee trough is formed in the westerlies.

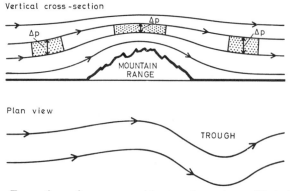

Fig. 8.19. Formation of an orographic trough or low. Variation of the pressure thickness (Δp) of an air column crossing a mountain range and the corresponding changes of curvature (assuming a uniform current with no change of shear).

8.7.3. *Long waves*

The systematic distribution of convergence and divergence in the atmosphere, shown in fig. 8.16, means that the mid-troposphere, 600 mb say, tends to be a level of non-divergence. At such a level, equation 8.14 simplifies to

$$\frac{\mathrm{d}}{\mathrm{d}t}(\zeta + f) = 0. \tag{8.15}$$

This equation states that air (which is not subject to divergence) will move in such a way that its absolute vorticity remains the same. Suppose that a uniform westerly current, expected on the basis of thermal wind theory, is disturbed (by topography, say) and turns somewhat polewards. Since the current then shares the Earth's greater cyclonic spin (increased f) its cyclonic spin relative to the Earth decreases until it becomes negative, i.e. anticyclonic. The air, having now turned equatorwards,

154

becomes subject to the opposite effect—decrease of f and algebraic increase of ζ—till it turns again polewards. So the current oscillates about its original latitude and comprises a series of ' long waves ', or ' Rossby waves ' (so-called after their discoverer), usually two to six in number round a circle of latitude (the smaller wave numbers in higher latitudes).

A formula that is derived for the speed of these waves shows that this is usually much less than that of the current as a whole and that, in certain conditions, the waves may be stationary or may even move westwards. In a given latitude, waves of a particular length (L_s) are stationary for a particular wind speed (U); if shorter than L_s, the waves move eastwards, and if longer than L_s they move westwards (in latitude 45°, for example, $L_s = 3000$ km when $U = 40$ kt); also, in a given latitude $L_s \propto \sqrt{U}$, i.e. increase of current speed drives the waves farther apart.

Some charts show the existence of long waves very clearly. Also, useful numerical predictions for 500 mb†, based on the conservation of absolute vorticity at that level, have been obtained. However, the characteristics of the long waves themselves are too difficult to disentangle from the complexities of the short waves for their movement to be determined on a day-to-day basis.

8.7.4. *Circulation indices: blocking*

One of the most striking features of the circulation of middle latitudes is the strong tendency for the alternation of periods when the flow is almost wholly zonal (west-to-east) with other periods when meridional flow (south-to-north and north-to-south in different longitudes) predominates. Usually, only a restricted range of longitudes is dominated by meridional flow but the effect is sometimes to be seen round a circle of latitude, at least in the northern hemisphere.

A usual measure of the surface circulation is the ' zonal index ' defined as the mean pressure difference between 35° and 55° latitude. This index has a well-marked seasonal variation (highest in winter, lowest in summer) but interest lies mainly in its variations over days or weeks. The change of synoptic type which is involved is often abrupt over a particular region: at other times there is a period when the flow is neither markedly of one type nor the other and which may well end with a reversion to the type which last prevailed. In any event, the phenomenon is of prime concern to a forecaster concerned with looking further ahead than 24 hours, say, and it has been closely studied.

The sequence of 1000–500 mb thickness charts in fig. 8.20 illustrates a change from high to low index. The thickness lines are initially (almost) west-to-east. A warm ridge and cold troughs are formed, and closed centres of high and low thickness then develop as indicated. Since surface centres of high and low pressure are also formed in almost

† 500 mb is the nearest standard level to 600 mb.

155

the same positions, the circulations tend to increase with height. The lows are called cut-off lows, since they are separated from the main low-pressure belt farther north; the high has a deep circulation which bars the normal eastward progression of depressions and is therefore called a blocking high.

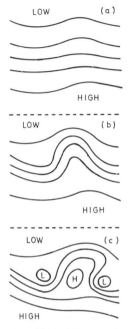

Fig. 8.20. Typical successive 1000–500 mb thickness patterns during the formation of cut-off lows and a blocking high.

In a blocking situation the weather changes little for days, or even weeks, over the region concerned. The characteristics depend—sometimes to rather critical limits—on how a particular region lies with respect to the blocking circulation. For example, if the axis of the blocking ridge is in the eastern Atlantic the British Isles is influenced by deep cold air carried from well north, and showery, unseasonably cold weather then prevails: on the other hand, if the axis lies between about 10°W and 10°E, fine and unseasonably warm weather is likely.

8.8. Surface analysis

8.8.1. General

Local weather is to a large extent dominated by air mass characteristics, and a notable change of weather usually involves the passage of a front and a consequent change from one air mass to another of markedly different properties.

156

Especially in middle latitudes, where air masses displace each other a few times per week on average (but may, exceptionally, do so as often as three times in one day), the appreciation of local weather is made much more alive by its interpretation in terms of the current synoptic situation. Press and television make this possible for the general public by reproducing an actual chart or a forecast chart or both. The ability to interpret these charts may be considerably enhanced by practice analysis, employing only material that is readily available. The information may be current, as in the B.B.C. Shipping Forecasts, or it may be one or two days old as in the official Daily Weather Reports; in the latter case there is the advantage of having the official solution readily available for direct comparison†.

Surface analysis consists of drawing isobars, and also of identifying air masses and locating fronts (air mass and frontal analysis). The former is an interpolative process based on individual readings; it is made easier by the isobar–wind relationship and presents little difficulty except near cols.

Apart from his specialized knowledge, the professional analyst has the advantage of being able to use the continuity between the successive surface charts which are available to him (he also has ready access to upper air information). His normal technique is, first, to estimate the positions of pressure centres and fronts by extrapolation from previous analyses and then to fix the actual positions by referring to the current observations. In certain cases—especially those of new development— the evidence presented by the latest chart concerning the existence or position of fronts may cause him to revise the analysis of the previous one or two charts.

A different approach is needed when a chart is analysed with no synoptic history (or upper air information) available. Initial rough drawing of isobars delineates the main pressure features and so indicates the more likely general positions of fronts. These obviously include well-marked troughs—though not all troughs are frontal. A depression with almost circular isobars and only small associated pressure tendencies is probably ' old '. It is unlikely to have a significant warm sector; its only likely front is a weak occlusion lying well to the east, probably showing cold front characteristics as it curves round towards the south of the depression and perhaps then linking with a depression farther west.

†(1) METMAPS, designed for the plotting and analysis of B.B.C. Shipping Forecast information, may be purchased from ' Weather ', Royal Meteorological Society, Cromwell House, High Street, Bracknell: an example is shown in fig. 8.21.
(2) Subscription copies of the Daily Weather Report may be obtained from the Director General, Meteorological Office, London Road, Bracknell, Berkshire: the synoptic codes are explained in the publication Met. 0.515 which may be purchased from H.M.S.O.; suitable charts (for example, Form 2216 on a scale $1 : 5 \times 10^6$) may be purchased from the same source.

On the other hand, a depression with large associated pressure tendencies very probably has a substantial warm sector, and warm and cold fronts are probably linked, through small or large ridges of high pressure, to preceding and following systems. It helps, in analysis of this kind, to have in mind the typical positions and linkages assumed by fronts in the various stages of development of a frontal depression.

These very general principles and also some further points to be made in the following section may serve as a guide to the positioning of fronts. The final process of accurate drawing of isobars sometimes indicates a need for slight adjustments of frontal positions which may be made if they are not in conflict with other evidence.

8.8.2. *Representativeness of observations*

The basic definition of a front is in terms of a change of temperature (T), but an accompanying change of dew point (T_d) is almost invariably involved. A front, then, should lie along (coincident) lines of maximum rates of change of T and T_d: in a sparse network of stations it may mark a line of actual discontinuity in these elements. Moreover, changes of wind velocity and pressure tendency are involved; and the interacting air masses have characteristic differences of weather, cloud type and amount, and visibility; finally, the fronts (usually) have associated belts of precipitation either in advance of, near, or behind them, depending on their type.

In principle, then, almost every element included in a synoptic plot provides evidence as to frontal position. With due allowance for errors of observation or transmission, the consensus of the observations should be overwhelming and the locating of fronts should provide no difficulty. This is indeed true of strong fronts in regions where the surface observations are representative of the air masses concerned. Difficulties become acute, however, when the fronts are weak and are located in regions where the observations are not representative.

The surface elements observed at sea are, in general, representative of their air mass, but various factors have nevertheless to be taken into account in the analysis of these observations. An obvious instance is the movement in winter of Pc air from Canada, say, out into the Atlantic, when very large T and T_d gradients are observed for hundreds of kilometres out from the coast. A gradient of this kind is not frontal in the true sense (it involves, for example, no wind discontinuity) but marks a zone of rapid transition of a Pc into a Pm air mass. In general, too, allowance has to be made for transitional effects on T and T_d in flow towards north or south. Other effects on T and T_d are non-latitudinal differences which occur in sea-surface temperature (a reported element in ship observations), and the tendency for precipitation to depress T and raise T_d towards the wet-bulb temperature of the air, by evaporation.

In most other respects, air mass and frontal analysis for sea regions is straightforward. In general, polar air masses are characterized by relatively large values of $(T - T_d)$, variable cumuliform cloud, showers, and good visibility; and tropical air masses by small $(T - T_d)$, large amounts of stratiform cloud, some drizzle, and poor visibility. Allowances are required, however, for the influence of curvature of isobars, as indicated in § 8.2.

The complicating factors mentioned above are equally relevant to the analysis of land observations or have their direct counterpart—for example the rapid transition of a warm air mass during its passage across a snow-covered region. Additional complications include the effects of topography, of the distribution of land and sea (or inland stretch of water), of local sources of atmospheric pollution (on visibility), and of marked diurnal variations of most of the elements. For example, a cold front which moves into the continent of Europe in winter may cause an *increase* of surface temperature when it displaces air which has been stagnant for some time and has cooled by radiation, while the converse is sometimes true of warm fronts in summer. The art of frontal analysis over land regions is to recognize, and as far as possible allow for, such and other effects on the observed elements.

8.8.3. *METMAPS*

Figure 8.21 is presented as a typical example of a surface analysis based on a B.B.C. Shipping Forecast.

(1) Reception and plotting

(*a*) Some practice is needed before a typical set of information can all be written at the speed at which it is broadcast. The form is designed to make this as easy as possible. Abbreviations are adopted and sea areas grouped as required.

(*b*) The coastal reports are plotted in the symbols shown on the chart inset. Note that the pressure is plotted in tens and units of millibars (the appropriate hundreds figure, 9 or 10, is never in doubt and is indicated on the numbered isobars). The wind speed is in Beaufort force, each half feather equivalent to one unit: the speed equivalents, in knots, of the Beaufort forces are indicated on the scale on the left of the chart.

(*c*) The winds predicted for the beginning of the forecast period are plotted in each sea area. If desired, predicted changes during the period may be added in the way shown on the chart inset.

(2) Analysis

(*a*) First, use may be made of the synoptic information which relates to the situation 5 hours earlier than the chart under discussion. Thus, the low in the north is assumed to have had about one-fifth of its predicted 24 hour motion; the trough is probably still in the sea areas Shannon to Humber; and a pressure of 1035 mb is assumed for the extreme south of the chart.

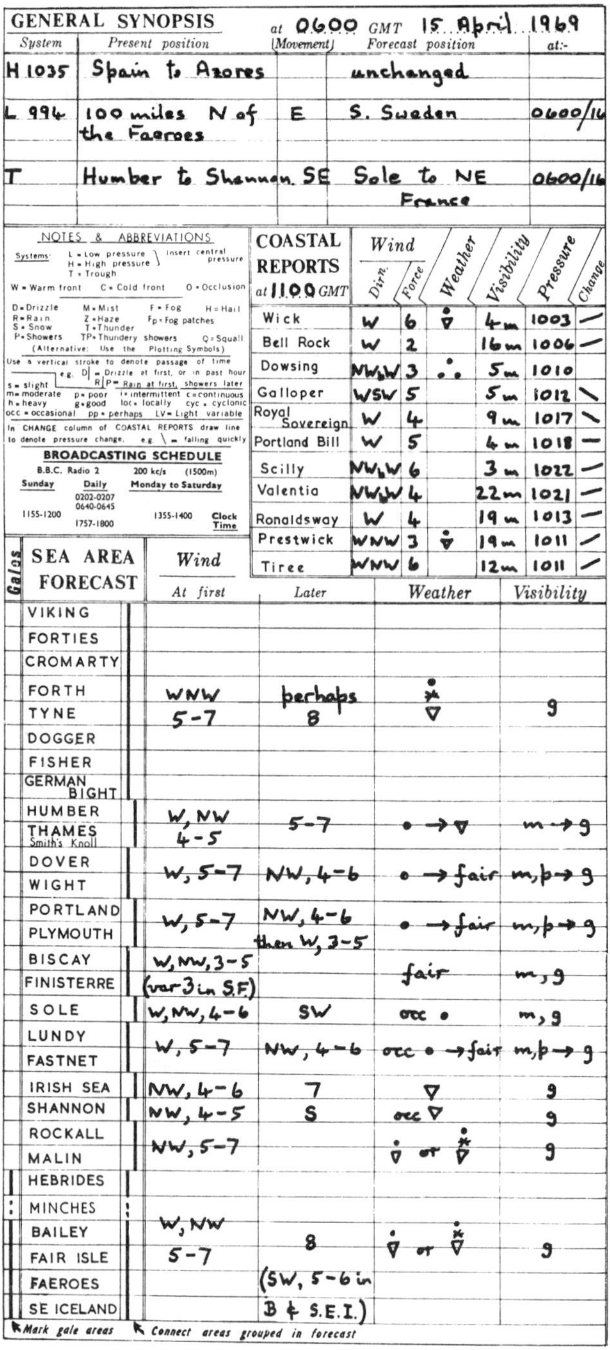

GENERAL SYNOPSIS at 0600 GMT 15 April 1969

System	Present position	(Movement)	Forecast position	at:-
H 1035	Spain to Azores		unchanged	
L 994	100 miles N of the Faeroes	E	S. Sweden	0600/16
T	Humber to Shannon.	SE	Sole to NE France	0600/16

NOTES & ABBREVIATIONS

Systems: L = Low pressure \ insert central pressure
H = High pressure /
T = Trough

W = Warm front   C = Cold front   O = Occlusion

D = Drizzle   M = Mist   F = Fog   H = Hail
R = Rain   Z = Haze   Fp = Fog patches
S = Snow   T = Thunder
P = Showers   TP = Thundery showers   Q = Squall
(Alternative: Use the Plotting Symbols)

Use a vertical stroke to denote passage of time
e.g. D| = Drizzle at first, or in past hour
|R|P = Rain at first, showers later

s = slight   D| = Drizzle at first, or in past hour
m = moderate   p = poor   i = intermittent   c = continuous
h = heavy   g = good   loc = locally   cyc = cyclonic
occ = occasional   pp = perhaps   LV = Light variable

In CHANGE column of COASTAL REPORTS draw line to denote pressure change.   e.g. \ = falling quickly

BROADCASTING SCHEDULE
B.B.C. Radio 2   200 kc/s   (1500m)

Sunday	Daily	Monday to Saturday	
1155-1200	0202-0207 / 0640-0645	1355-1400	Clock Time
	1757-1800		

COASTAL REPORTS at 1100 GMT

	Wind Dir.	Force	Weather	Visibility	Pressure	Change
Wick	W	6	▽̇	4m	1003	/
Bell Rock	W	2		16m	1006	/
Dowsing	NW,W	3	∴	5m	1010	
Galloper	WSW	5		5m	1012	\
Royal Sovereign	W	4		9m	1017	\
Portland Bill	W	5		4m	1018	—
Scilly	NW,W	6		3m	1022	/
Valentia	NW,W	4		22m	1021	/
Ronaldsway	W	4		19m	1013	/
Prestwick	WNW	3	▽̇	19m	1011	/
Tiree	WNW	6		12m	1011	/

SEA AREA FORECAST

Gales	SEA AREA	Wind At first	Later	Weather	Visibility
	VIKING				
	FORTIES				
	CROMARTY				
	FORTH	WNW	perhaps	✳̇ / ▽̇	9
	TYNE	5-7	8		
	DOGGER				
	FISHER				
	GERMAN BIGHT				
	HUMBER	W, NW	5-7	• → ▽	m → 9
	THAMES (Smith's Knoll)	4-5			
	DOVER	W, 5-7	NW, 4-6	• → fair	m, p → 9
	WIGHT				
	PORTLAND	W, 5-7	NW, 4-6 then W, 3-5	• → fair	m, p → 9
	PLYMOUTH				
	BISCAY	W, NW, 3-5 (var 3 in S.E.)		fair	m, 9
	FINISTERRE				
	SOLE	W, NW, 4-6	SW	occ •	m, 9
	LUNDY	W, 5-7	NW, 4-6	occ • → fair	m, p → 9
	FASTNET				
	IRISH SEA	NW, 4-6	7	▽	9
	SHANNON	NW, 4-5	S	occ ▽	9
	ROCKALL	NW, 5-7		▽̇ or ✳̇/▽̇	9
	MALIN				
	HEBRIDES				
:	MINCHES :				
	BAILEY	W, NW	8	▽̇ or ✳̇/▽̇	9
	FAIR ISLE	5-7			
	FAEROES		(SW, 5-6 in		
	SE ICELAND		B & S.E.I.)		

↖ Mark gale areas   ↖ Connect areas grouped in forecast

Fig. 8.21.
METMAP for
15 April, 1969.

160

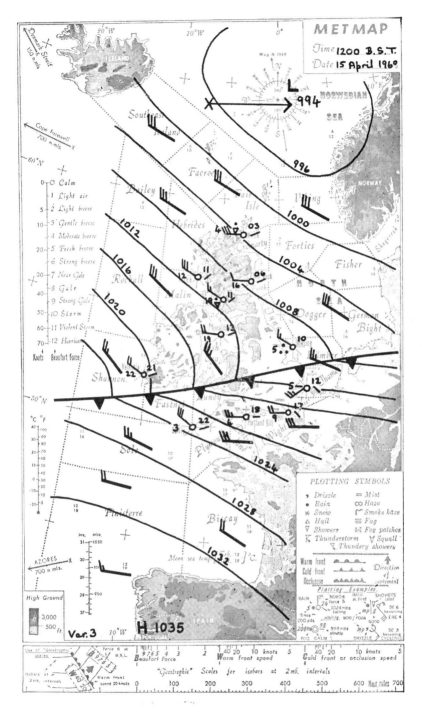

(*b*) Isobars are drawn† on the basis of the (predicted) positions of the pressure centres, together with the winds in the sea areas and the winds and pressures in the coastal reports. A 4 mb interval is used in this instance (each value exactly divisible by 4): with lighter gradients a 2 mb interval may be preferred.

(*c*) In the absence of T, T_d information the frontal trough is placed on the basis of the wind shift, pressure tendency, associated weather and relative ease of drawing the isobars near the trough. The relationship of this particular trough to the isobars and to the low in the north strongly suggests that it is a cold front, and it is marked as such.

(The southward passage of this front is, in fact, being slowed down by a small wave depression which is moving quickly east along the front without deepening. The continuous moderate rain reported at Dowsing, and the opening-out of the isobars on the front, are the only indications of this. More detailed observations at this time show a considerable amount of rain and a weak circulation round the wave located near the Wash.)

(*d*) Frontal speed and Beaufort force at any point may be determined by the use of the scales shown under the chart.

Questions on Chapter 8

8.1. A cold front lies west-to-east in middle latitudes in the northern hemisphere. The wind direction in the cold air is 060° while the distance apart of the isobars in the direction normal to themselves is 4 mb/100 km. Calculate the geostrophic wind speed in the cold air and the speed of the front.

8.2. If the (equilibrium) slope of the frontal surface in question 8.1 is 1 : 60 and the temperatures of the cold and warm air are $-2°C$ and $+6°C$, what is the geostrophic wind velocity in the warm air?

8.3. Suppose, in fig. 8.17 d, that r is 250 nautical miles and that the corresponding wind speed is 50 kt. What is the vorticity at distance r from the centre? What is the wind shear required to maintain constant vorticity inwards from r and what is the magnitude of this vorticity? Verify by calculation that the condition $Vr = $ constant in the region outwards from r makes the vorticity there zero.

8.4. An observed quantity, such as surface temperature, changes at a fixed place at a rate which depends on its rate of advection (owing to the motion of air in the direction of a gradient—see § 6.5.4) and on its rate of change within an individual moving element. For temperature, the local change is the partial derivative $(\partial T/\partial t)$, the individual element change is the total derivative (dT/dt), the

† A fundamental principle in drawing an isobar is that the associated wind direction cannot reverse at some point along it. Thus if the motion is initially downwind in tracing an isobar it must remain so whatever the shape of the isobar. If, in a particular drawing, this is not true then the pressure field is not correctly represented.

advective change is $-V(\partial T/\partial s)$ (where $\partial T/\partial s$ is the rate of temperature rise in the downwind directions) and thus

$$\frac{\partial T}{\partial t} = \frac{\mathrm{d}T}{\mathrm{d}t} - V\left(\frac{\partial T}{\partial s}\right).$$

A uniform north-westerly airstream flows at 24 kt from south Greenland, where the temperature is $-10°C$, to north Scotland, at a (downwind) distance of 1000 nautical miles, where the temperature is $+5°C$. If surface heating warms the air by 1°C per 3 hours during its passage what is the rate of temperature change in north Scotland?

8.5. The rate of steady rainfall, over a given area, is mainly controlled by the rate of convergence of air at low levels and by the depth and water content of the converging air. Find the rate of rainfall for an average inward velocity of 5 m s^{-1}, extending through a layer of depth 4 km across a ' circular ' area of radius 100 km, the average water content of the converging air being 8 g m^{-3}. (In a steady state all the water entering the area is assumed to fall out there, as in § 5.3.)

Show that, for a constant rate of inflow, the rainfall rate decreases in proportion to increase of radius of the area affected.

CHAPTER 9

micrometeorology

MICROMETEOROLOGY may be defined as the detailed study of physical phenomena in the very lowest layers of the atmosphere. Because it is within these regions that we live and work, and within which most of our food-stuffs originate, it is considered most appropriate in a book of this kind to include a fairly detailed introduction to this particular branch of meteorology. Moreover, with minimum outlay and only a little trouble, a single student or group of students may carry out many illuminating experiments within the realm of micrometeorology.

9.1. *The nature of airflow near the ground*

At heights in excess of 500 m or so above the Earth's surface, horizontal air motions proceed in a largely geostrophic manner. Such motions are essentially unretarded by friction and in consequence are generally smooth and free from turbulence. However, below 500 m (the upper limit of the ' planetary boundary layer ', or ' friction layer ') turbulence is much in evidence, together with other manifestations of the frictional drag exerted on air motion by the Earth's surface—e.g. the progressive backing of the wind observed between the top and bottom of the friction layer. Between a level of 50 m or so and the surface, the speed of the wind (which decreases relatively little between 500 m and 50 m) reduces more and more rapidly towards zero. This sub-region is often termed the ' surface boundary layer '—the particular character of the wind therein being determined largely by the physical nature of the underlying surface, but also by the magnitude and sign of the vertical gradient in temperature. The region between 500 m and 50 m is in effect a zone of transition between the smooth geostrophic flow in the free atmosphere and flow of an essentially turbulent nature near the ground.

9.1.1. *Wind speeds over a uniform level surface*

If anemometers are erected at several heights (z) above any reasonably uniform and sufficiently extensive level area and the observed mean wind speeds ($u(z)$) plotted against z, the resulting wind ' profile ' is found to have a shape similar to that shown in fig. 9.1; i.e. the vertical wind shear ($\partial u/\partial z$) is found to be largest near the surface itself and to decrease

164

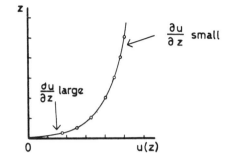

Fig. 9.1. Typical wind profile over a uniform surface.

progressively upwards. Moreover, provided winds are strong enough to counter effects of vertical temperature gradient (see § 9.3.3), plotting of $\partial u/\partial z$ against $1/z$ invariably produces a straight line relationship; so that in general

$$\frac{\partial u}{\partial z} = A \cdot \frac{1}{z},\qquad(9.1)$$

where the parameter A, although independent of z, is a function of wind speed and of the nature of the surface in question. On integration of equation (9.1),

$$u(z) = A \ln z + B,\qquad(9.2)$$

where B is the appropriate constant of integration. This relationship is of the form found, in the laboratory, to describe the shape of the wind profile in a fully developed turbulent boundary layer. Thus the nature and characteristics of airflow close to the Earth's surface can be described and explained in terms of existing boundary-layer theory, which has been adequately confirmed by controlled laboratory experiments. The following sub-sections provide a brief introduction to turbulent-boundary-layer theory, somewhat over-simplified perhaps but at least sufficient to provide a plausible derivation of the logarithmic profile equation (equation (9.2)) in complete form.

9.1.2. *Flow within a fluid boundary layer*

Fluid moving over a level surface exerts a horizontal force on the surface in the direction of motion of the fluid; such a drag force is usually expressed per unit area of surface and termed shearing stress, τ. Conversely, the surface exerts an equal and opposite retarding force on the fluid: this force does not act on the bulk of the fluid (at least in the first instance) but only on its lower boundary and on a region of more or less restricted extent immediately above, known as the fluid boundary layer. Flow within such a layer may be purely laminar, a condition seldom if ever found over extensive natural surfaces, or essentially

165

turbulent with a laminar sub-layer close to the surface itself†—fig. 9.2. The shearing stress exerted on a surface by fluid flow is generated within the boundary layer and transmitted downwards to the surface in the form of a momentum flux. (Dimensions of shearing stress can be expressed MLT^{-2}/L^2, force per unit area; or MLT^{-1}/L^2T, momentum per unit area per unit time.)

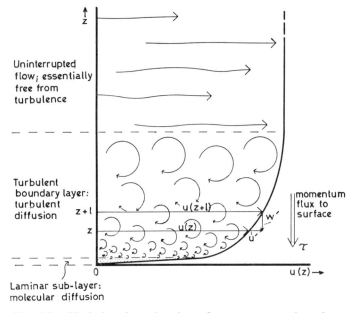

Fig. 9.2. Turbulent-boundary-layer flow over a smooth surface.

This downward flux of streamwise momentum arises from the sheared nature of the flow within the boundary layer, and derives from inter-action between this shear and random (vertical) motions within the fluid. In the laminar sub-layer these diffusive motions are entirely molecular in origin, character, and scale; whereas in the turbulent zone they are macroscopic—in scale at least—discrete 'lumps' of fluid being displaced by turbulent action through a characteristic distance, known as the 'mixing length' l (analogous perhaps to molecular mean free path), before merging with the surrounding fluid.

9.1.3. *Shearing stress via the mixing length concept*

Referring to fig. 9.2; suppose that a lump of fluid originally at the level $(z+l)$ and having the appropriate mean velocity $u(z+l)$ is dis-placed to the level z by the action of turbulence; the instantaneous

† But see § 9.2.1.

velocity at z then exceeds the mean value by an amount u' given by $u(z+l)-u(z)$; i.e., to a first approximation,

$$u' = l(\partial u/\partial z) \qquad (9.3)$$

The subsequent merging of this lump of fluid with its surroundings results in a quantity of momentum $\rho u'$ per unit volume being contributed to the flow at the level z. Moreover, if the magnitude of the transient vertical velocity imparted to the lump of fluid is w' then the rate at which momentum is conveyed downwards across unit horizontal area by such a motion must be $\rho u'w'$. Assuming that a constant momentum flux of this magnitude is communicated by like process to the top of the laminar sub-layer, whence by molecular means to the surface itself, we may write

$$\tau = \rho u'w'. \qquad (9.4)$$

9.1.4. *The friction velocity u_**

It is convenient, however, to express shearing stress in terms of the ' friction velocity ' u_*, such that

$$\tau = \rho u_*{}^2, \qquad (9.5)$$

where u_*, like the product $u'w'$, is constant throughout a region of constant momentum flux, or shearing stress, τ. If we assume that u' and w' are merely *comparable* in size we may deduce that u_* is representative of the magnitude of the velocity fluctuations in turbulent-boundary-layer flow. It is however justifiable to assume *equality* of u' and w', so that

$$u' = w' = u_*, \qquad (9.6)$$

and to substitute u_* for u' in equation (9.3). The task remains of finding an expression for the mixing length l in equation (9.3).

9.1.5. *Interpretation of the mixing length concept*

The apparently chaotic motion of fluid in a turbulent boundary layer may be visualized as a smooth mean flow on which large numbers of eddies are superimposed. Each eddy moves with the mean flow velocity $u(z)$, to which its own internal motions, which we may identify with the components u' and w'†, are added to give the instantaneous velocity at any given point. It is with the scale of these individual eddies that mixing length can be identified. Intuitively, we expect this scale to decrease downwards through the boundary layer (as depicted in fig. 9.2) until, at the surface itself, all turbulent motions are inhibited and $l=0$. (Technically, all eddy motion ceases at the top of the laminar sub-layer, within which l is effectively the molecular mean free path.) The simplest possible deduction from this reasoning is that l is directly

---

† And therefore with u_*, which is sometimes called ' eddy velocity '.

proportional to distance above the surface—and this is confirmed by experiment; so that

$$l = kz. \tag{9.7}$$

Moreover, the constant of proportionality, k, is found to be independent of the nature of the underlying surface: for historic reasons k is known as von Kármán's constant; it has the value 0·40.

9.1.6. *The wind profile equation in complete form*

From equations (9.3), (9.6) and (9.7) the parameter A in equation (9.1) can be equated to u_*/k; i.e.

$$\frac{\partial u}{\partial z} = \frac{u_*}{kz}, \tag{9.8}$$

and, in place of equation (9.2),

$$u(z) = \frac{u_*}{k} \ln z + B. \tag{9.9}$$

Equation (9.9) describes the shape of the wind profile in turbulent-boundary-layer flow down to the level of the laminar sub-layer, within which region it is physically incapable of doing so—witness the wind speed which it predicts at $z = 0$, namely $u(0) = -\infty$. This shortcoming is avoided in practice by restricting the zone of application of equation (9.9) to the region above a level z_0, where z_0 is defined by the requirement that $u(z_0) = 0$. Equation (9.9) then takes the practical form

$$u(z) = \frac{u_*}{k} \ln \left(\frac{z}{z_0} \right), \tag{9.10}$$

in which z_0 includes the role of constant of integration previously held by B. The length z_0 does however have a certain physical significance, as we shall see in the following section.

9.2. *The influence of surface roughness on the wind*

9.2.1. *Roughness in the aerodynamic sense*

The size of the length z_0 is found to depend on the ' roughness ' of the surface over which measurements are made. It should be understood however that by roughness in this context is meant roughness in the aerodynamic sense which may differ to a greater or lesser extent from one's own conception of the nature of a particular surface, obtained by sight or touch. For example, a moderate sea is no more rough, aerodynamically speaking, than a football field; moreover, a surface which appears rough to the senses will be ' aerodynamically smooth ' if the depth of the laminar sub-layer (a decreasing function of u_*) is sufficient to exclude the roughness elements of the surface from the turbulent flow above. For all aerodynamically smooth surfaces z_0 has the same small value, of order 10^{-2} mm—some 100 times less in fact than the depth of the laminar sub-layer.

In nature, however, most surfaces are ' aerodynamically rough '; i.e. the individual roughness elements, such as blades of grass, penetrate far enough into the region of turbulent flow to ensure that the shearing stress on the surface is made up of aerodynamic drag forces acting on each one of them individually, rather than on an all-enveloping laminar sub-layer.   For such aerodynamically rough surfaces the size of z_0 depends entirely on the character of the individual surface elements—not only on their size, but also on their shape and on their distribution (spacing, orientation, etc.): typical values of the ' roughness length ' z_0 are given in Table 9.1.

Type of surface	Typical roughness length, z_0 mm
Smooth mud flats, or ice	0·01
Smooth snow (on short grass)	0·05
Blown sand	0·5
Blown snow } Mown lawn	1·0
Short grass	5
Rough pasture	10
Heather moor	25
Long grass and most tall field crops	50 to 100
Forest	perhaps 500

Table 9.1.   Roughness lengths of various surfaces.

9.2.2. Roughness in relation to shearing stress and mean wind speed

It follows from equation (9.10) that the greater the roughness of a surface—specified by z_0—the less is the wind required (at a given height) to generate a particular shearing stress—specified by u_*; or, conversely, the greater is the shearing stress generated by a particular wind. These conclusions are best represented graphically, in the following manner.   Equation (9.10) is first written in the form

$$\ln z = \frac{k}{u_*} u(z) + \ln z_0. \qquad (9.11)$$

This shows that when $u(z)$, as the abscissa, is plotted against $\ln z$, as the ordinate, a straight line of slope k/u_* is obtained, which crosses the $\ln (z)$ axis $(u(z) = 0)$ at the point $\ln z = \ln z_0$.   Figure 9.3 a shows the ' logarithmic wind profiles ' corresponding to a friction velocity of 0·5 m s^{-1} (i.e. $\tau = 0·3$ N m^{-2}) over surfaces with roughness lengths 0·1, 1·0, 10 and 100 mm: the wind speed indicated at $z = 2$ m decreases from

169

12·3 m s⁻¹ with $z_0 = 0.1$ mm to 3·7 m s⁻¹ with $z_0 = 100$ mm.   Figure 9.3 b, on the other hand, corresponds to a fixed wind speed of 7·5 m s⁻¹ at $z = 2$ m and shows that the shearing stress increases from 0·1 N m⁻² ($u_* = 0.3$ m s⁻¹), over the surface with $z_0 = 0.1$ mm, to 1·2 N m⁻² ($u_* = 1.0$ m s⁻¹) over that with $z_0 = 100$ mm.   Included in figs. 9.3 a and 9.3 b are sketches of the actual wind profiles involved; these portray the actual conditions more realistically than the logarithmic representation.

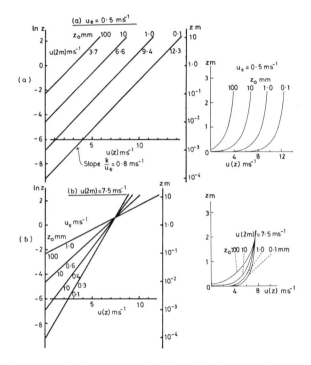

Fig. 9.3.   Logarithmic and actual wind profiles for several values of z_0.
(a) $u_* = $ constant $= 0.5$ m s⁻¹; (b) $u(2$ m$) = $ constant $= 7.5$ m s⁻¹.

A further point of some relevance may be deduced from fig. 9.3 b: namely that the rougher a surface is then the smaller in general will be the mean wind speed close to it—although the turbulence or gustiness (specified by u_*) will be greater.   This agrees with the figures quoted in § 6.4, which imply for a given isobar spacing (i.e. for a given geostrophic wind) that the wind speed at 10 m over the sea ($z_0 \sim 0.1$ to 1·0 mm) is on average approximately twice that at the same level over land ($z_0 \sim 10$ to 100 mm).   Also, as the destructive force of wind on an isolated obstruction (house, barn, tree, etc.) is proportional (roughly) to the square of the wind speed, this conclusion emphasizes the need to

170

maintain the overall roughness of a countryside liable to the ravishes of wind, even if this means replanting hedgerows and tree-plantations previously uprooted in the name of economy.

9.2.3. *The drag coefficient C_D*

So far we have specified surface roughness in terms of the parameter z_0. It is frequently preferable, however, to relate roughness to the magnitude of the shearing stress (i.e. drag force per unit horizontal area) generated by a given amount of wind, a relation best expressed by the surface drag coefficient C_D, defined by

$$\tau = \rho u^2(z)\, C_D(z). \tag{9.12}$$

(Clearly, the ' given amount of wind ' depends on the height at which it is measured, so that C_D also must be a function of z. In practice C_D is referred to a standard height, usually 2 m or 10 m.) On comparing equations (9.12) and (9.5) a simple relationship is obtained, namely:

$$C_D(z) = \left(\frac{u_*}{u(z)}\right)^2. \tag{9.13}$$

Because much of the drag on natural rough surfaces (vegetation) is due to the action of pressure forces on individual surface elements (leaves, stems, etc.), and because such forces vary as (wind speed)^2, it may be deduced from the form of equation (9.12) that the value of C_D for a particular surface must be independent of wind speed[†]. It follows from equation (9.13) that, at any particular height above such a surface, the ratio between the friction velocity and the mean flow velocity must be independent of wind speed; while from equation (9.10) we may conclude that a constant drag coefficient implies a constant roughness length. Thus the value of C_D (or of z_0) obtained at any reasonable wind speed for a particular surface can be deemed to be the unique value corresponding to that surface, and therefore can be used to calculate the drag on it at any other wind speed (see problem 9.2).

9.2.4. *C_D as a transfer coefficient*

To the micrometeorologist, the coefficient C_D is more than just an aerodynamic drag coefficient, larger values of which imply greater surface roughness and vice versa. It is a coefficient the size of which provides a measure of the effectiveness with which momentum is transferred (downwards) across a turbulent boundary layer and onto the surface below. (Note that, for a given wind speed, a large drag coefficient implies (via equation (9.12)) a large shearing stress, which in

[†] This is not true at wind speeds sufficiently high to promote streamlining (see Problem 9.3): nor is it quite true at very low wind speeds, because of viscous effects.

turn implies a large downward flux of momentum.) It follows that C_D must relate in some manner to the 'conductance' of the boundary layer as a whole, a conductance deriving for the most part from the diffusive action of turbulence. This important property of turbulence is treated in § 9.3.

9.2.5. *Effect of a change in surface roughness*

Finally it should be emphasized that, for airflow to adjust to the roughness of a particular surface throughout any appreciable depth, a considerable downstream distance is required—as illustrated in fig. 9.4. Figure 9.4 *a* shows wind blowing from a surface of roughness length z_{01} (assumed to extend indefinitely upwind, beyond point A) over a significantly rougher surface ($z_0 = z_{02}$) extending downwind from O, at least as far as B. The wind profile characteristic of the first surface,

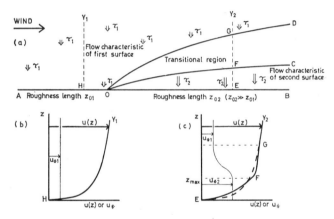

Fig. 9.4. The effect of a change in surface roughness from z_{01} to z_{02}, $z_{02} \gg z_{01}$. (*a*) Extent of influence of new surface roughness; (*b*) wind and friction velocity profiles incident on the new surface; (*c*) wind and friction velocity profiles downstream from the roughness change.

and incident on the second at O, is sketched in fig. 9.4 *b*, together with portrayal of a friction velocity (u_{*1}) independent of z—this being indicative of a constant shearing stress throughout the boundary layer, a stress (τ_1) equal to that on the first surface. The curve OD in fig. 9.4 *a* represents the upper boundary of the region within which the shape of the incident wind profile is modified in any way by the change in surface roughness: curve OC, on the other hand, indicates the more limited vertical extent to which the incident profile is fully adjusted to this change: the common sector DOC must be regarded as transitional.

172

These statements are better illustrated in fig. 9.4 c which shows the wind profile typical of a point E at some distance downwind from the change in surface roughness. The incident wind profile is indicated also†, together with the vertical distribution of shearing stress, represented by u^*. Clearly, the transitional zone is the region in which shearing stress is modified‡ from its initial value to that (τ_2) characteristic of the new surface roughness.

The depth (EF) to which the modified wind profile is characteristic of the underlying surface increases only slowly with distance downwind from the change in roughness. This is frequently a matter of practical concern because, in nature, homogeneous surfaces tend to be of limited extent. As a rule of thumb the wind profile at a distance X downwind from a change in surface roughness will be characteristic of that surface up to the level

$$z_{max} = \frac{X}{40} \qquad (9.14)$$

and no further.

In any field experiment designed to measure roughness length, for example, anemometers must be confined to the region, below z_{max}, in which the flow is 'fully developed'.

9.3. *Vertical transport by turbulence*

" The feature of the surface wind of which the micrometeorologist is most conscious is not its magnitude and direction but its turbulence. ... The main effect of turbulence is to cause enhanced diffusion of matter, heat and momentum." O. G. Sutton.

Turbulence in any fluid has a diffusive action many orders of magnitude more effective than that arising from the random movements of molecules. For example, liquid dye introduced carefully into the bottom of a beaker of water (at room temperature) requires several hours to diffuse uniformly throughout the beaker, whereas the same result is produced in seconds merely by stirring. On an atmospheric scale, molecular diffusion is completely insignificant: indeed, without the mixing action engendered by turbulence in the planetary boundary layer, conditions at and near the surface of the Earth would differ catastrophically from those with which we are familiar. The smoke at present pouring from densely populated and industrial areas would no longer mix thoroughly into the atmosphere, from which it is subsequently washed out by rain. Instead, such pollution would most likely remain within a shallow layer close to the surface, and because evaporation from the oceans (itself a product of turbulent exchange)

† Note: conservation of mass (flow) requires that the areas between the wind speed curves and the ordinate axis be equal; hence the local acceleration in flow below G.

‡ In this example the friction velocity is increased from u_{*1} to u_{*2} by the relatively large wind shear present in the lower part of the transitional zone.

would be practically at a standstill no rain would be available to wash it out; consequently it would probably accumulate and thicken indefinitely†. In addition, any part of the solid globe not permanently enshrouded by mist, fog, or smog would probably suffer enormous diurnal fluctuations in temperature owing to the lack of an effective heat-transfer mechanism.

Clearly, then, the turbulence which is generally present in the planetary boundary layer endows the airflow in these regions with the necessary element of 'conductivity' required to maintain essential vertical fluxes of water vapour, sensible heat and pollution. In deriving equations for such fluxes, it is both simple and instructive to follow up the use of electrical terminology (such as conductivity), and employ an analogue technique, as follows.

9.3.1. *Flux equations; use of electrical analogy*

In electricity, where Ohm's law applies to a wire, the relation between its resistance, the current through it, and the potential difference across it is

$$\text{resistance} = \text{potential difference/current}. \qquad (9.15)$$

Ohm's law itself began life as an analogy with heat flow. We can now reverse the course of history. If we write Ohm's law as

electrical resistance = electrical potential difference/rate of charge flow

the thermal equivalent is,

thermal resistance = temperature difference/rate of heat flow

or thermal resistance = temperature difference/heat flux.

Here, the analogues are obvious; and we can extend this kind of equation to many other cases of flow, with suitable quantities replacing potential difference, and rate of flow—which we shall call flux when it means the rate of flow across unit area. The analogue equation is

$$\text{'resistance'} = \text{'potential difference'/'flux'} \qquad (9.16)$$

Equation (9.16) may be applied without further modification to problems involving the turbulent flux of any property (vertically) across a region of boundary-layer flow, provided that the appropriate 'potential' is used—namely the concentration, i.e. quantity per unit volume, of the property concerned. (The direction of such a flux is from a region of high to a region of low concentration.) It follows that turbulent resistance is defined by the quotient (concentration difference ÷ flux). As this is equivalent to (difference in quantity per unit volume) ÷ (quantity per unit area per unit time), we see that whatever the dimensions of the property being transferred, say Q, those of turbulent resistance are $(QL^{-3}) \div (QL^{-2}T^{-1})$, i.e. TL^{-1}, or (velocity)$^{-1}$.

Associated directly with the turbulence in a fully developed boundary

† At present the total mass of solid pollutants in the atmosphere is estimated to be 10 000 000 tons.

layer is a downward flux of momentum of magnitude τ (see § 9.1.2). The potential difference driving this flux between any appropriate level z and the surface is equal to the corresponding difference in momentum per unit volume, namely $\rho u(z) - \rho u_0$ or, as the wind speed u_0 on the surface itself is everywhere zero, simply $\rho u(z)$. It follows from equation (9.16) that the resistance to the transfer of momentum between z and the surface, written as $r_D(z)$, is given by $\rho u(z)/\tau$. However, $\tau = \rho u_*^2$ (equation (9.5)), so that

$$r_D(z) = \frac{u(z)}{u_*^2}. \tag{9.17}$$

The precise form of the relationship suggested in § 9.2.4 between the momentum transfer coefficient $C_D(z)$ and the corresponding boundary-layer conductance, is found by combining equations (9.17) and (9.13) to give

$$r_D(z) = \frac{1}{u(z)\,C_D(z)}. \tag{9.18}$$

Thus $(r_D(z))^{-1}$, the conductance of the boundary layer (with regard to momentum) between z and the surface, is equal to the product of $u(z)$ and the coefficient $C_D(z)$.

For convective transfer of heat, the appropriate potential is heat content per unit volume, i.e. $\rho c_p T$. Thus the resistance offered to a heat flux of magnitude H between a surface whose average temperature is T_0 and a level z at which the mean temperature is $T(z)$ must be given by

$$r_H(z) = \frac{\rho c_p (T_0 - T(z))}{H}. \tag{9.19}$$

The analogous expression for an evaporative flux E is

$$r_V(z) = \frac{\rho}{\frac{8}{5}p} \cdot \frac{(e_0 - e(z))}{E}, \tag{9.20}$$

in which the appropriate potential ρ_v is expressed in terms of vapour pressure—see § 2.3.1.

The resistances $r_H(z)$ and $r_V(z)$, as defined by the preceding equations, can be calculated under any circumstances from a knowledge of the size of the appropriate flux and the corresponding potential difference. On the other hand, use of these equations to determine the fluxes themselves requires that a value for $r_H(z)$ or $r_V(z)$ be found; and although the resistance $r_D(z)$ may be substituted for either of $r_H(z)$ or $r_V(z)$, this is frequently a poor approximation because of the different transfer processes involved at the surface itself—especially when it is aerodynamically rough[†]. As a result of this, and also because of

† Because the drag on (momentum transfer to) such a surface is due predominantly to pressure forces acting on individual roughness elements, while matter and heat must rely on molecular diffusion alone to cross the laminar boundary layer enveloping each element, both $r_H(z)$ and $r_V(z)$ are often significantly greater than $r_D(z)$.

difficulties in measuring (or even defining) mean conditions of temperature or vapour pressure on a vegetated surface, flux equations for heat and water vapour are usually expressed in the following forms:

$$H = \frac{\rho c_p (T(z_1) - T(z_2))}{r_H(z_1, z_2)} \qquad (9.21)$$

and

$$E = \frac{\rho}{\frac{8}{5}p} \cdot \frac{(e(z_1) - e(z_2))}{r_V(z_1, z_2)}. \qquad (9.22)$$

z_1 and z_2 are two levels, z_1 being the nearer to the surface, while each $r(z_1, z_2)$ is the appropriate resistance to transfer between z_1 and z_2. Except at the surface itself the vertical transport of any one property, e.g. water vapour, is brought about by the identical turbulent motions which result in the simultaneous vertical transport of any other—such as momentum.   Consequently, each property must experience the same turbulent resistance, so that

$$r_H(z_1, z_2) = r_V(z_1, z_2) = r_D(z_1, z_2), \qquad (9.23)$$

where, by deduction from equation (9.17),

$$r_D(z_1, z_2) = \frac{u(z_2) - u(z_1)}{u_*^2}. \qquad (9.24)$$

9.3.2. *Heat flux and other calculations*

Let us suppose that the observations of mean wind speed, temperature and vapour pressure quoted in Table 9.2 were made in a region of fully developed boundary-layer flow (i.e. at least 100 m downwind from the nearest small obstacle or significant change in surface roughness) and that they are not significantly influenced by effects of free convection (see § 9.3.3).

z m	$u(z)$ m s^{-1}	$T(z)$°C	$e(z)$ mb
2	4·8	19·8	18·7
1	4·0	20·5	19·3

Table 9.2.

The associated vertical fluxes of sensible heat and water vapour may then be determined, as follows.

(i) Before the appropriate turbulent resistance is derived from equation (9.24), a value for u_* must first be found.   From equation (9.10),

$$\frac{u(z_1)}{u_*} = \frac{1}{k} \ln \left(\frac{z_1}{z_0} \right) \quad \text{and} \quad \frac{u(z_2)}{u_*} = \frac{1}{k} \ln \left(\frac{z_2}{z_0} \right),$$

so that

$$u_* = k(u(z_2) - u(z_1))/\ln \left(\frac{z_2}{z_1}\right), \qquad (9.25)$$

which, from Table 9.2 and with $k = 0.40$, gives

$$u_* = 0.46 \text{ m s}^{-1}.$$

(ii) From equation (9.24),

$$r_D(z_1, z_2) = 3.8 \text{ s m}^{-1}.$$

(iii) This resistance value is substituted into equation (9.21), via equation (9.23), together with $\rho = 1.2$ kg m^{-3} and $c_p = 10^3$ J kg^{-1}, to give the sensible heat flux

$$H = 0.22 \text{ kW m}^{-2};$$

and (iv), into equation (9.22), together with $p = 1000$ mb, to give the water vapour flux, or evaporation rate,

$$E = 1.2 \times 10^{-4} \text{ kg m}^{-2} \text{ s}^{-1}.$$

In practical units this last is equivalent to a rate of evaporation of 0.43 mm of water per hour; while in terms of the latent heat involved it is equivalent to an energy flux

$$L_v E = 0.30 \text{ kW m}^{-2}.$$

The total flux of energy leaving the surface by the mechanism of turbulent exchange is equal to the sum of H and $L_v E$, i.e. 0.52 kW m^{-2}. This in turn may be equated to the net flux of radiant energy R_{net} incident on the surface at the time, provided that the corresponding flux of energy into the surface by molecular conduction may be neglected (and also that mean conditions are essentially steady-state).

The ratio between the sensible and latent heat fluxes leaving the surface, known as Bowen's ratio B and given by $H/L_v E$, is equal to 0.73—typical of a vegetated surface. At a water surface, in comparison, the partition of R_{net} into sensible and latent heat is much less equal and values of B around 0.1 are common†.

Some confirmation of the surface type is obtained by calculating its roughness length z_0. This is done most conveniently by dividing equation (9.10), with $z = z_1$, by equation (9.25) to give the ratio

$$x = \frac{u(z_1)}{u(z_2) - u(z_1)} = \frac{\ln \left(\frac{z_1}{z_0}\right)}{\ln \left(\frac{z_2}{z_1}\right)}. \qquad (9.26)$$

† B has the value -1 at a water surface when $R_{net} = 0$, while large positive values are typical of arid desert regions.

From Table 9.2, $x = 5$, so that z_0, given by the relation

$$z_0 = \frac{z_1}{\left(\dfrac{z_2}{z_1}\right)^x},$$ (9.27)

is 33 mm—a value typical of grass of moderate length.

Finally, first approximations to the mean temperature and vapour pressure at the surface itself are obtained by substituting $r_D(z)$ for $r_H(z)$ and $r_V(z)$ in equations (9.19) and (9.20), as follows.

The resistance $r_D(z_1)$ is found from equation (9.17), with $u_* = 0.46$ m s^{-1}, to be,

$$r_D(z_1) = 19.0 \text{ s m}^{-1},$$

which, in equation (9.19) with $H = 220$ W m^{-2} and $T(z_1) = 20.5°C$, gives

$$T_0 = 24.0°C,$$

and, in equation (9.20), with $E = 1.2 \times 10^{-4}$ (kg m^{-2} s^{-1}) and $e(z_1) = 19.3$ mb, gives

$$e_0 = 22.3 \text{ mb.}$$

Note the relatively large temperature difference, of order 4 K, which is maintained between the surface itself and the region 1 to 2 m. This is typical of clear daytime conditions at most times of the year, but is especially beneficial to the microclimate of the surface regions during the spring growing period.

9.3.3. Vertical temperature gradients in relation to turbulent exchange

The logarithmic wind profile equation (equation (9.10)) and those relations which derive from it, such as equation (9.25), apply rigorously (in the theoretical sense) only in conditions of neutral stability. Moreover, it is known that under strong lapse or inversion conditions equation (9.10) fails to describe the shape of the observed wind profiles, while the equality expressed by equation (9.23) is likewise found to fail under such conditions. Consequently, it is the main purpose of this section to establish the extent to which conditions may in fact depart from strict neutrality without such a departure invalidating the use of equations (9.10) and (9.23)—in effect, without such a departure causing significant modifications in the simple scheme of turbulent exchange associated with boundary-layer flow under conditions of neutral stability.

Figure 9.5 shows typical profiles of velocity, temperature and density in a turbulent boundary layer within which temperature decreases with height. Two reference levels (z_1 and z_2) are chosen, within this boundary layer, such that $z_2 = z_1 + l$, where l is the local mixing length.

Consider a lump of fluid of volume V. Its total kinetic energy is made up of two independent components: one derives from its horizontal progress as part of the mean flow, while the other is associated with its own random movements as part of the forced turbulence present in the boundary layer. Accordingly, the kinetic energy (K.E.) associated with the purely vertical movements of this lump of fluid is represented by the product $\frac{1}{2}\rho V w'^2$. However, from equations (9.6) and (9.3),

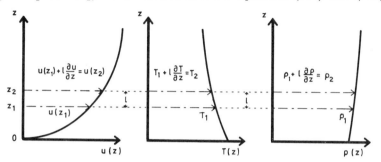

Fig. 9.5. Profiles of wind speed, temperature and density in a turbulent boundary layer. Turbulent exchange of each property occurs through the same characteristic distance, or mixing length, l.

$w' = u' = l\ \partial u/\partial z$, so that this vertical component of the 'natural' kinetic energy of turbulence, $(K.E.)_n$, can be written in the form

$$(K.E.)_n = \tfrac{1}{2}\rho V l^2 \left(\frac{\partial u}{\partial z}\right)^2. \tag{9.28}$$

Now, in the course of turbulent exchange, such a lump of fluid, initially at the level z_1 and with density ρ_1 appropriate to that level, is likely to find itself at the level z_2 in surroundings of density ρ_2. Consequently, at this latter level, it is subject to a positive buoyancy force of magnitude $g(\rho_2 - \rho_1) V = F$, say. If it is assumed that the buoyancy force acting on the lump of fluid during its motion from z_1 to z_2 increases uniformly (with distance above z_1) from zero at z_1 to F at z_2† then the work done (W.D.) by buoyant action in the course of this motion may be equated to the product of the force $\frac{1}{2}F$ and the distance $(z_2 - z_1)$. That is

$$\text{W.D.} = \tfrac{1}{2}g(\rho_2 - \rho_1)(z_2 - z_1)\ V. \tag{9.29}$$

This amount of work goes directly to increase the upward kinetic energy of the lump of fluid, equal to $(K.E.)_n$, by an amount $(K.E.)_b$, to which latter it may be equated. Thus

$$(K.E.)_b = \tfrac{1}{2}g(\rho_2 - \rho_1)(z_2 - z_1)\ V, \tag{9.30}$$

† Equivalent to assuming a linear temperature profile locally between z_1 and z_2.

which, on substitution of l for $(z_2 - z_1)$, $l\dfrac{\partial \rho}{\partial z}$ for $(\rho_2 - \rho_1)$ and finally† of $-\dfrac{\rho}{T} \cdot \dfrac{\partial T}{\partial z}$ for $\dfrac{\partial \rho}{\partial z}$, can be written in the form most convenient for the present analysis, namely

$$(\text{K.E.})_b = -\tfrac{1}{2}\rho V l^2 \cdot \frac{g}{T} \cdot \frac{\partial T}{\partial z}. \tag{9.31}$$

Clearly, the size of the ratio $(\text{K.E.})_b/(\text{K.E.})_n$ relates to the relative importance of free and forced convection in determining the structure of a fluid boundary layer (shape of velocity profile, nature of turbulence, etc.). In practice, the parameter $-(\text{K.E.})_b/(\text{K.E.})_n$, known as Richardson's number Ri, is used as a measure of instability or stability in a region of boundary-layer flow. From equations (9.28) and (9.31),

$$\text{Ri} = \frac{g}{T} \cdot \frac{\dfrac{\partial T}{\partial z}}{\left(\dfrac{\partial u}{\partial z}\right)^2}. \tag{9.32}$$

Thus in lapse conditions Ri is negative, and in inversion conditions positive.

It may be deduced that in a region where Ri exceeds the value $+1$, spontaneous turbulent motions are impossible; likewise, any eddy motions penetrating such a region from elsewhere must rapidly die out. In practice, a positive Richardson's number of only a small fraction of unity, say $+0.2$, is sufficient to cause extensive damping of turbulence in the surface boundary layer, and so produce essentially laminar conditions of flow. We note, then, that turbulent exchange is strongly inhibited by the presence of a temperature inversion—such as is produced by radiative cooling of the ground under clear skies.

At the other extreme, in a region where Ri is algebraically less than -1, say, free convection replaces forced convection as the dominant transfer mechanism, thereby generating the greatly enhanced turbulent exchange which is characteristic of strong lapse conditions. Even for Ri as small as -0.1 the influence of the associated temperature lapse is enough to amplify significantly the turbulence already in the boundary layer and, as a result, to alter the shape of the logarithmic wind profile.

The range of Richardson's number within which equations (9.10) and (9.23) can in fact be considered to be valid is small indeed, namely:

$$-0.01 < \text{Ri} < 0.01. \tag{9.33}$$

Outwith this range, account must be taken of the effect of buoyancy on the shape of the mean wind profile and also on the relative rates at which

† By differentiating equation (2.5).

momentum, mass and heat are transported across the turbulent boundary layer.

In conclusion, let us check on our supposition that the observations quoted in Table 9.2 are not significantly influenced by effects of free convection. Equation (9.32) is first simplified by equating g to ~ 10 m s^{-2} and T to 288 K, and is then written in the finite difference form

$$\text{Ri} \doteqdot 0 \cdot 035 \frac{\Delta T \cdot \Delta z}{(\Delta u)^2}. \qquad (9.34)$$

From Table 9.2: $\Delta T = -0 \cdot 7$ K, $\Delta z = 1 \cdot 0$ m, and $\Delta u = 0 \cdot 8$ m s^{-1}; so that $\text{Ri} = -0 \cdot 04$, in reasonable accord with the limits set by equation (9.33).

Questions on Chapter 9

9.1. Wind speeds measured at $z_1 = 2$ m and $z_2 = 10$ m above an extensive grassy surface are $u(z_1) = 4 \cdot 5$ m s^{-1} and $u(z_2) = 6 \cdot 5$ m s^{-1}. Determine graphically or otherwise the roughness length of the surface.

9.2. Show that $C_D(z) = k^2/(\ln{(z/z_0)})^2$, and find $C_D(z_2)$ and τ for the preceding example. If at some other time $u(z_2) = 4 \cdot 2$ m s^{-1} what is τ?

9.3. During a period of strong winds $u(z_1) = 15$ m s^{-1} and $u(z_2) = 20$ m s^{-1}, over the same grassy surface as in 9.1. Show that streamlining of the grass blades must take place at such high wind speeds. Again, find the shearing stress τ.

9.4. The following observations were made above a uniformly vegetated surface of sufficient extent.

Level	z m	$u(z)$ m s^{-1}	$T(z)°$C	$e(z)$ mb	CO$_2$ concentration mg m^{-3}
z_2	2	5·8	19·3	14·4	540
z_1	1	4·9	19·9	15·3	530

Before calculating z_0, u_* and τ confirm that Ri is within the limits specified by equation (9.33). Determine the sensible heat flux (H) and the flux of water vapour (E). Express E as a rate of evaporation (properly, transpiration) in units of mm of water per hour. Calculate the latent heat flux associated with E, and the corresponding value of Bowen's ratio. Show that the vegetation is assimilating carbon dioxide at the rate of 10·8 grammes per hour per square metre of ground surface. Finally, obtain an estimate of the mean surface values of temperature, vapour pressure, relative humidity and CO$_2$ concentration.

181

9.5. Shearing stress within a laminar boundary layer is conveniently written $\tau = \rho\nu(\partial u/\partial z)$, where ν, the molecular diffusivity of momentum (or kinematic viscosity), is independent of position (z) and equal to $1{\cdot}5 \times 10^{-5}$ m^2 s^{-1}.

Show that turbulent shearing stress may be expressed in the analogous form $\tau = \rho K_M(z)(\partial u/\partial z)$, where $K_M(z)$, the turbulent diffusivity of momentum (or eddy viscosity), is given by the product ku_*z. (Clearly then, $K_M(z)$ increases with z.) Use the relevant data in the preceding example to show that turbulent diffusivity at a height of 10 m is some five orders of magnitude greater than molecular diffusivity.

Show that, in terms of the electrical analogy (§ 9.3.1), $K_M^{-1}(z)$ has dimensions of resistivity. Confirm this by integrating $K_M^{-1}(z)\,dz$ between two levels z_1 and z_2, to give $r_D(z_1, z_2)$.

9.6. Demonstrate from equations (9.21), (9.22) and (9.24) that, if equation (9.23) holds, profiles of wind speed, temperature and humidity must be similar in shape.

CHAPTER 10
The general circulation

DIFFERENT from the problems of day-to-day or even shorter-period variations of the elements, over more or less limited regions, with which this book is almost exclusively concerned, is the problem of the atmosphere considered as a single, though vast, heat engine operating over an extended period of time. These larger scales of space and time naturally include the smaller but are very much more complex.

The general circulation poses the most fundamental problem in meteorology. In a sense it comprises the whole of the subject although this is a definition which is too wide to be helpful. Its fundamental nature has long been recognized and significant attempts were made to explain it at least as long ago as the early 18th century. One such explanation, by Hadley in 1735, though restricted in its applicability (as he himself recognized) is still considered relevant to an important part of the circulation.

Like all old problems in science that of the general circulation is difficult. It was no doubt once thought that when observations increased in their range and number an explanation of the circulation would result. The reverse has rather been the case in that each extension of observations over the surface of the globe, and also upwards from the surface, has tended to raise new problems faster than it has solved existing ones.

There is a restricted sense in which the problem may be considered no longer to exist, in that it has been shown to be possible (first in the U.S.A.) to apply the differential equations of heating and motion to an unrealistic atmosphere (isothermal and motionless) and transfer it by integration, step by step, into one with features of wind, pressure and temperature, in the troposphere and stratosphere, very like those that are observed in the real atmosphere. The calculations involved would be quite impracticable if a large electronic computer were not available. Such research work, still in its early stages, is beyond the scope of this book. Meteorologists believe that it has very great potential and that it now offers perhaps the only way—certainly the most likely one—of making further progress. The problem remains at present in that even the most recent work does not provide a complete explanation, in physical terms, of the main features of the general circulation or of the interactions of its various parts.

In this chapter we outline the genesis of the circulation and some of its complications, discuss the results of standard observations and of

183

experiment in the laboratory, present the cellular circulation model that best explains the observations, and finally show how the general circulation determines the distribution of climate over the Earth's surface.

10.1. *General characteristics*

10.1.1. *Genesis and interactions*

The source of the atmosphere's energy is solar radiation which, having different intensities in different latitudes, causes atmospheric temperature differences and therefore pressure differences (see the discussion of thermal winds in Chapter 6). The global circulation patterns that are established by the pressure fields at different levels have many regular features, especially when averages are taken over a longish interval of time—some years, say—and these features comprise the general circulation.

In principle it may appear possible, on the basis of thermal winds, for the circulation established by the radiation field over a uniform surface to be purely horizontal and along lines of latitude. In that event, equatorial latitudes would be raised to a very high surface temperature at which the intensity of long-wave radiation emitted by the ground would balance the solar radiation absorbed while, in contrast, the surface radiative equilibrium temperature in polar latitudes would be very low. But this is not the real situation for at least two reasons†. First, vigorous convective motion would be established in the low-latitude air on account of its very high surface temperature (which is to say that pure radiative equilibrium conditions cannot hold when there is strong surface heating). Second, the Earth's surface is far from uniform and this implies that the gradient of temperature is not strictly north–south, so that the (thermal wind) motion has, in general, a meridional (north–south) component. Because of the convergence of the meridians in high latitudes and the need to maintain continuity of mass it is impossible to have purely horizontal motion in such circumstances. Thus we see that both local and large-scale vertical motions are inevitable and we infer that the general circulation must also be three-dimensional.

The complexities of atmospheric circulation on a global scale arise in part from sheer size but a more fundamental difficulty lies in the absence of simple cause and effect between different parts of the system, owing to multiple cross-linking between the various processes. The following summary of the main processes will indicate what is meant.

The variations of solar radiation with latitude and season, at the atmosphere's outer fringe, are determined by geometrical astronomy and are precisely known. The fractions of solar energy which are reflected to space, or absorbed in the atmosphere, or reach the Earth's surface depend on the path lengths of the radiation through the atmosphere, on

† A more important but less obvious reason is given in §10.3.3.

the amounts of water vapour and dust traversed, and on cloud amounts. The fraction absorbed at a particular surface depends on the local albedo. The temperature response of the surface to a given absorption of energy is very different for land and sea. Radiation exchanges between Earth and atmosphere, and also the evaporation rate, are determined by the local meteorological elements. Each time water changes phase in the atmosphere a large amount of energy is absorbed or released.

The winds redistribute warm and cold air and also water vapour, and this redistribution in turn affects the winds. The exchange processes at the Earth's surface (involving heat, water vapour and momentum) and the wind structure close to it depend on surface roughness which is very variable over the land and which is also a function of wind speed for air-flow over the sea or over vegetation. The motion of the air is barred by mountains which, by causing water vapour to condense, act as high-level sources of energy. Finally—perhaps, for the long term, the most imponderable factor of all—the atmosphere continuously interchanges energy with the oceans which have a heat capacity about one thousand times greater than that of the atmosphere†.

10.1.2. *Time fluctuations*

In many parts of the world and for periods of days, weeks, or even months the atmosphere slips, as it were, out of its normal mode of circulation into some other pattern: the regions concerned then experience unseasonal warmth or cold, dryness or wetness. Such appearances and disappearances of circulation anomalies, as they are called, are an intrinsic part of the atmosphere's functioning and have to be explained.

The longer the interval considered the smaller are the anomalies of mean wind, rainfall, temperature etc., compared to the long-term averages. However, in some regions, small imbalances of a given sign (in temperature, say) may occur in significantly more years than not and then comprise a change of climate in the regions concerned. Most of these fluctuations are minor and temporary but they have included, on a geological time scale, various Ice Ages as well as the much longer periods when there has been no permanent ice anywhere on the Earth's surface. In all such cases changes in the general circulation are implied.

10.2. *Observations*

10.2.1. *Time- and space-averaging*

One of the principal features of synoptic charts for the Earth as a whole is the much greater constancy of surface pressure systems, and therefore of winds, in the tropics and sub-tropics compared to higher latitudes.

† They cover about 70% of the Earth's surface and have an average depth of about 4 km.

The circulation in low latitudes is dominated by vast currents of air—the trade winds—which move equatorwards from semi-permanent belts of anticyclones in the sub-tropics. Middle and higher latitudes are, to a much greater extent, affected by short-lived and fast-moving pressure systems, with corresponding fluctuations in the circulation patterns. Figure 10.1 is intended to illustrate a synoptic chart over one half of the earth's surface for an individual occasion near an equinox†: at this time

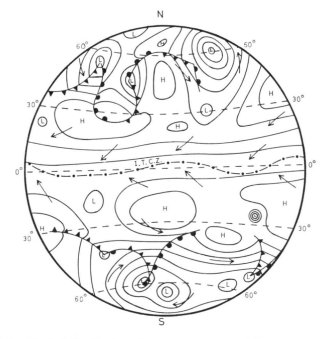

Fig. 10.1. Typical distribution of pressure systems and fronts over one half of the Earth's surface near an equinox; surface flow is indicated by arrows.

of the year the intertropical convergence zone (ITCZ) at which the circulations of the two hemispheres meet is not far from the equator.

If averages are calculated from a large number of occasions the variable pressure features of individual charts are reduced in importance and the mean distributions of sea-level pressure and winds may finally approximate to those shown in fig. 10.2 (which apply, strictly, to a hypothetical uniform surface and not to the real surface of the Earth).

† The flow on the poleward side of 30°N in fig. 10.1, though realistic, is less zonal (i.e. west–east) than is usual even for this hemisphere.

In the summer hemisphere the greatest daily total of solar radiation reaches the earth's surface between latitudes 20° and 30° and the smallest daily total received is at the pole; however, the latitudinal gradient of received radiation is relatively small because the lower peak intensity (local noon) values received in high latitudes are to some extent compensated by their long hours of sunshine in this season. In the winter hemisphere the maximum radiation input is at 0° and the latitudinal gradient of received radiation is large up to 60° because both peak intensity and hours of sunshine decrease with distance from the equator; polewards of 60° the total radiation reaching the Earth's surface in the winter hemisphere is very small. Since, basically, the atmosphere

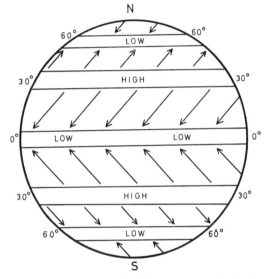

Fig. 10.2. Idealized distribution of surface pressure and winds over the Earth.

is driven by the latitudinal gradient of the heat which it receives we must expect seasonal changes in the circulation: therefore average values of pressure, etc. are calculated for separate seasons, or separate calendar months, over a period of years, as below.

(1) Surface observations

The distributions of mean-sea-level pressure and wind, for January and July, are shown in figs 10.3 and 10.4. The main features are as follows:

(a) The ITCZ is farther north in July than in January (these are the extreme north and south positions reached throughout the year).

(b) The high and low pressure systems favour certain longitudes rather than latitudes 30° and 60° in general; thus we have the Aleutian and Icelandic lows and the Pacific, Atlantic and (south) Indian Ocean highs.

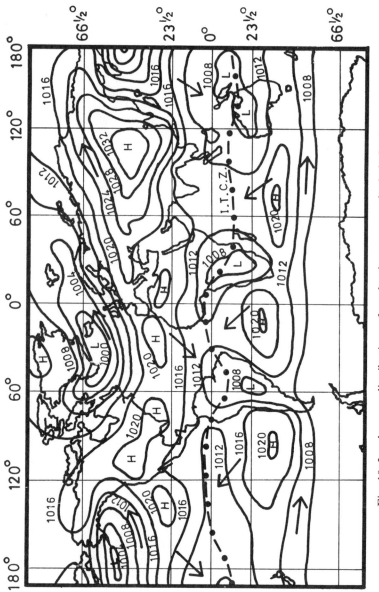

Fig. 10.3. Average distributions of sea-level pressure and winds for January.

188

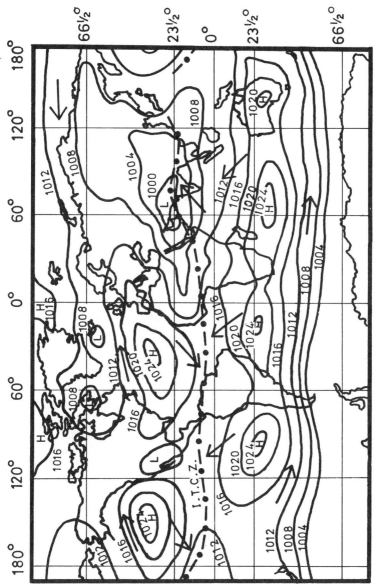

Fig. 10.4. Average distributions of sea-level pressure and winds for July.

189

(c) The sub-tropical high pressure belts are about 5° nearer the equator in winter than in summer and the mid-latitude westerlies are therefore broadest in the winter season: they are also at their strongest then.

(d) The continents of North America and Asia are often dominated in winter by anticyclones which are formed there by prolonged radiational cooling. These features, and also the large heat low formed over N.W. India in summer, are complexities of pressure and circulation which are caused by the distribution of land and sea in the northern hemisphere. This complexity is less conspicuous in the southern hemisphere where pressure and wind approximate much more closely to the idealized distribution of fig. 10.2.

(2) Upper atmosphere

Figures 10.5 a and b show how nearly zonal the mean circulation (known as the circumpolar vortex) is†, and also show the increase in its strength from summer to winter. Because of the preponderance of zonal flow at all levels the diagram shown in fig. 10.6 contains most of the factual information about the mean circulation, up to at least the tropopause, for the two hemispheres and both seasons. Note for example, the increase of the sub-tropical westerlies to above 40 m s⁻¹ in winter; note also the related movement of the upper ' equatorial easterlies ' northwards and southwards with the sun.

With increasing height in the stratosphere, the winds become entirely seasonal—westerly in winter, easterly in summer—and reach very high values near the stratopause. At still higher levels, wind speeds decrease and finally reverse. Wind velocity and temperature distribution are related by thermal wind theory just as at lower levels. Vertical components of motion are linked with the seasonal changes; they account, for example, for the much higher temperature at the mesopause in winter than in summer, as mentioned in § 2.2.3.

Although the process of averaging is useful in reducing day-to-day complexities, so revealing fundamental aspects of atmospheric circulation, it must not be forgotten that the variability which is averaged out (mainly the mobile pressure systems of middle and higher latitudes) plays a vital role in the transfer of heat and other properties from one latitude to another: figs 8.2 and 8.18 illustrate how effective these transfers are. It is therefore implied that the general circulation cannot be satisfactorily explained in terms of the mean motions alone.

10.2.2. *Tracers*

Much has been learned about cyclones and anticyclones by studying the horizontal and vertical components of motion of air parcels round and through them. The method involves the tracking of air parcels which have been labelled by a particular value of a property such as

† The standing waves visible in the patterns are considered to be caused mainly by the obstruction to the flow presented by the Rocky Mountains.

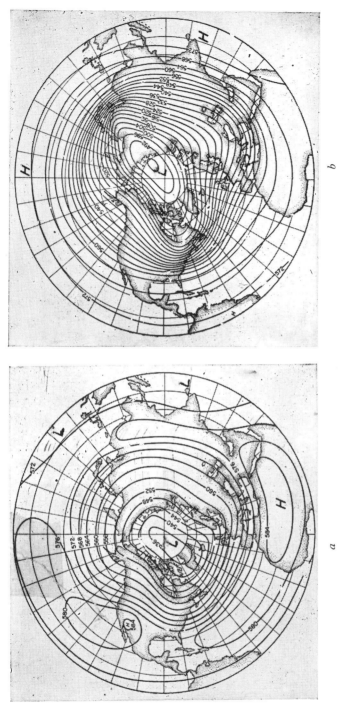

a

b

Fig. 10.5. Mean 500 mb contours for northern hemisphere for (*a*) July and (*b*) January (after Scherhag). The mean wind blows parallel to the contours in a counterclockwise direction and with a speed that varies inversely with both the contour spacing and sine of the latitude.

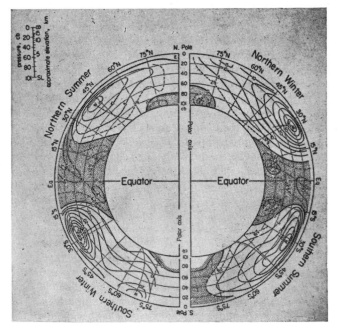

Fig. 10.6. Variation of zonal mean speed (m s⁻¹) with latitude, height and season over the Earth (after Mintz). Speed is constant along any contour and easterlies are shaded.

mixing ratio or wet bulb potential temperature. The use of tracers of this kind cannot readily be extended to studies of the general circulation because the air is subject, sooner or later, to processes such as precipitation, evaporation and radiation for which the identifying property is not conservative. There are, nevertheless, certain quantities which act as effective tracers of global atmospheric motion, as follows.

(1) Frost point† measurements

Except near the mesopause, which is at a height of about 80 km, the coldest part of the atmosphere is the equatorial tropopause at about 18 km where temperatures are around $-80°C$. This must obviously represent the upper limit for the frost point of air which passes through this region because, when air is cooled to this temperature, any excess moisture (relative to a frost point of $-80°C$) is removed by condensation. Within the troposphere or the lower stratosphere, air which has a measured frost point as low as about $-80°C$ is considered to have been dried out at the equatorial tropopause. Stratospheric air in middle and high latitudes sometimes has such low frost points and it is inferred that

† Temperature of saturation with respect to ice; it is given by ZY on fig. 2.4.

192

the air concerned has ascended to the lower stratosphere at the equator, losing moisture in the process, and has then moved polewards to middle and high latitudes. This very dry air usually makes its return to the troposphere in these latitudes, sometimes through the tropopause breaks which are associated with active fronts: this is also true of atmospheric ozone, a more generally useful tracer element, which we now consider.

(2) Ozone variations

Ozone is formed at high levels by the action of a particular component of sunlight and its maximum rate of formation is therefore reached near the equator where sunlight is strongest. Similarly, at any given latitude its rate of formation is greatest in the summer months. Although ozone is destroyed by a different component of sunlight the expectation is that

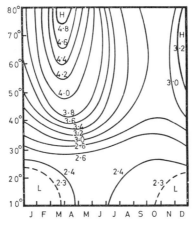

Fig. 10.7. Distribution of ozone with latitude and season over the Northern Hemisphere (after Dobson). The units are millimetre thickness of pure ozone at normal (surface) temperature and pressure.

equilibrium should be established between the photochemical processes of formation and destruction at a higher level of concentration of the gas when the sunlight is strong than when it is weak: in that event the maximum concentration of the gas would occur near the equator and, in any latitude, during summer. Figure 10.7 shows how the mean concentration of ozone does, in fact, vary in time and space in the northern hemisphere. Similar results, based on fewer data, are available for the southern hemisphere. In both hemispheres the ozone concentrations are largest in high latitudes and in spring and are smallest in low latitudes and in autumn.

The explanation of these observed ozone distributions is that only a small fraction of the ozone which is contained in an atmospheric column is in the photochemical equilibrium region. Most of the gas is in the lower stratosphere (below about 30 km) having been transported there, not by turbulent mixing which is almost non-existent in this stable region, but by large-scale descent of the air with which it is mixed. The ozone in the lower stratosphere is well protected, by the ozone above it, from the destructive component of solar radiation. On the other hand, when the stratospheric air, relatively rich in ozone, moves down into the troposphere it mixes very much more readily in the vertical; it then fairly soon reaches the Earth's surface where the ozone which it contains is quickly destroyed, by contact with vegetation for example. It is thus only the ozone which is between the tropopause and the level of about 30 km that is semi-permanent and able to act as a tracer. The mass of air that can be accommodated between these levels is much greater in high latitudes than in low (in the ratio of their tropopause pressures— see figs 2.1 and 2.2): the higher latitudes therefore have the greater capacity for storing ozone.

The following features of global circulation are implied by the variations shown in fig. 10.7. The stratospheric air of high latitudes cools by net radiation throughout the winter and the mass of dense, cold air so formed attains a maximum rate of subsidence in early spring†. This air moves down from the ozone formation region in the high strato- sphere and becomes concentrated between the tropopause and about 30 km in high latitudes: the measured ozone in a vertical column is then a maximum. On the other hand, upward motion near the equator occurs in the troposphere (which contains very little ozone) and con- tinues into the lower stratosphere: ozone amounts are therefore small near the equator. The equatorial air spreads polewards in the strato- sphere and subsequently returns to the troposphere in middle and high latitudes.

The horizontal and vertical transfers of ozone in middle and higher latitudes are not of a steady nature but are connected with the occurrence of depressions and anticyclones, the ozone transfer being in this respect similar to the transfer of heat and other quantities. Day-to-day fluctuations of measured ozone amount in these latitudes are at least as big as the seasonal changes. Though connected in a general way with depressions and anticyclones they are most closely associated with the changes at the tropopause and in the lower stratosphere.

(3) Radioactive tracers

The radioactive particles produced in the testing of nuclear weapons in the atmosphere act as effective tracers of motion during the period that

† Because of the very stable lapse rates that prevail in the stratosphere, subsidence in this region causes large rises of temperature at given heights or pressure levels (see Problem 5.5): this type of occurrence, which may be quite sudden, is a feature of later winter or spring, and has been called ' explosive warming '.

they remain in the atmosphere. This period varies a great deal depending mainly on the height but also on the latitude of the explosion.

Particles which fail to reach the stratosphere but are so small as to have negligible speeds of fall have been observed (by air sampling, for example) to circle the globe several times and to range over a moderately wide latitude band. They are removed within weeks, mainly by being washed out during precipitation. Particles which reach the stratosphere have a life history which agrees well with that of ozone (with due allowance for their different origins). Thus, although most of the testing has been carried out in low latitudes the subsequent fallout of the particles has been mainly in middle and high latitudes, with a maximum in spring—further evidence of an upward then poleward circulation at the equator. From the few occasions of nuclear bursts in the mesosphere it seems that the particles may remain suspended there for years rather than for months: if this is so then the circulation of the mesosphere would appear to be more isolated from that of the stratosphere than is the case with the stratospheric and tropospheric circulations.

10.3. *Experiment and theory*

10.3.1. *The rotating vessel experiment*

A cylindrical vessel about 30 cm across, with a flat bottom and open at the top, contains a few centimetres depth of water: the vessel is heated at its rim and cooled at its centre and is also made to rotate about a vertical axis passing through its centre. This simultaneous heating and cooling produces a temperature difference between the water near the rim and that near the centre. In terms of the Earth, the water corresponds to the atmosphere, the heated rim is the equator, the cooled centre is a pole and the period of revolution (a fraction of a minute) corresponds to a day. The object of this so-called dishpan experiment is to determine the resulting motion of the water as shown, for example, by aluminium filings or powder spread on the surface; to see in what respects it resembles the motion of the atmosphere; and, especially, to find out how the motion is affected by changes in the rates of rotation and heating. The motion viewed from a stationary point outside the apparatus corresponds to the motion of the Earth's atmosphere viewed from space; while a moving film taken by a camera which rotates with the cylinder corresponds to atmospheric motion as viewed on the Earth.

The most interesting results of the experiment are probably those which concern the changing character of the motion as the rotation speed is changed. If, first, heating and cooling are applied without rotation then the introduction of a dye shows that the surface flow is inwards towards the ' pole ' and that the flow near the bottom of the vessel is outwards towards the ' equator ': the motion is cellular, with upward motion at the heat source and downward motion at the heat sink. Suppose now that the cylinder is rotated slowly in a counter-clockwise

195

direction (reckoned west-to-east). Then the surface water, moving polewards, acquires a westerly component of velocity relative to the vessel and this tends to increase towards the centre; and the water at the bottom of the vessel moves equatorwards and with an easterly component. As the speed of rotation is increased this steady motion gradually changes into one which is characterized by long wave-like patterns on which smaller vortices are superimposed (see Plates 14 and 15). Finally, with very fast rotation (one revolution in a few seconds), the motion breaks up into a large number of very small and quickly-changing rotation elements superimposed on a general rotation of the water at the same speed as the cylinder.

The experimental model of course differs from the atmosphere in a number of fundamental ways one of which relates to the rotation rate. For the atmosphere, the rotation about the local vertical decreases from a maximum at the pole to zero at the equator (see §6.1.2): while for the water in the vessel the rotation rate is the same at all 'latitudes'. Thus the results observed at the slowest rotation speed of the cylinder would be expected to apply, if at all, to the equatorial regions; and those obtained at the faster rotation speeds may apply to higher latitudes. In both respects the experimental results agree well with what is observed in the atmosphere. Thus, as we saw earlier, the evidence from routine observations and from tracers is that the air ascends at the equator, then moves polewards at higher levels acquiring an increasing westerly component as it does so; also, that surface air moves towards the equator and with an easterly component: these agree with the experimental observations at low speeds. In higher latitudes the normal field of atmospheric motion is a combination of long waves and smaller superimposed disturbances, similar to what is observed in the vessel over a wide range of rotation speeds. There is no atmospheric counterpart, however, for what is observed at the fastest experimental speeds of rotation.

10.3.2. *Conservation principles*

(1) Energy and water

Since the basic conservation laws of physics (those relating to energy, mass, momentum etc.) are involved in the solution of nearly all types of physical problem the meteorologist is much concerned with how these principles apply to the atmosphere. An example of this was discussed in Chapter 3, in relation to fig. 3.4, where we assumed that, since the net radiation for the Earth and atmosphere as a whole is zero in a steady state, then excesses of radiation in one latitude must balance deficits in another. We may further assume, with reference to fig. 3.4, that a deficit, say, of radiation in a particular latitude must be compensated there by an influx of energy, by other means, of just the right amount to keep the temperature steady: this condition applies to the Earth and atmosphere separately, as well as jointly, and applies also to each individual place in a given latitude.

The budgets that are drawn up—those for energy and water, in particular—are complex and rely on data which, in many cases, are sparse and inadequate. However, various cross-checks may be applied as a useful means of improving accuracy. For example, the water budget must be consistent with the overall energy budget since the latent heat exchanges involve large amounts of energy. Again, the transport of water in the atmosphere across different circles of latitude, say, may be calculated from wind and humidity data for upper levels (only the meridional components of velocity are important in such a case) and the values of flux divergence of water vapour between the various parallels of latitude may thus be obtained. Suppose, for example, that this divergence is positive in a particular latitude belt, i.e. more water vapour is transported from the area concerned than enters it. This must mean that the air gains more water by evaporation and transpiration from the region than it loses to it by precipitation. The sign and magnitude of the calculated divergence should therefore check with the difference between the separate estimates of evaporation and precipitation over the area concerned.

The various measurements show that evaporation greatly exceeds precipitation between about 20° and 40° latitude and is less than precipitation in all other latitudes, but especially near the equator (fig. 10.8). Thus in both hemispheres, the winds carry moisture equatorwards and polewards from about latitude 30°. The local balance of water, which must hold in the long term, is achieved by net flow within the oceans towards latitude 30° from both north and south. These facts concerning moisture transport, based on latitudinal means, conceal important regional differences which are caused by the uneven distribution of land and sea, more especially of mountain ranges which are so important in their influence on the distribution of precipitation.

(2) Momentum

For the Earth-atmosphere system taken as a whole, angular momentum remains constant with time because no external turning force acts about the axis of rotation (principle of the conservation of angular momentum). However, easterly surface winds extract angular momentum from the Earth's surface (the Earth's rotation being from west to east) and westerly surface winds add angular momentum to the surface, in both cases through the frictional force which acts between the air and the ground in the direction of the surface wind. In the steady state, in which the Earth's speed of rotation remains constant, the average torque exerted by westerly surface winds must balance that exerted by easterly surface winds. This, then, is a condition that must be satisfied by the general circulation. It is of interest to note that accurate measurements indicate that the length of day is systematically very slightly greater (by about 2×10^{-3} s) in February than in August owing to a corresponding variation of winds on the global scale. Since these are internal changes and involve no force external to the whole system, the total angular momentum of earth and atmosphere is unaffected: it follows that the momentum change of the atmosphere from February to August must be equal and opposite to that of the Earth. as indicated by the changing day length.

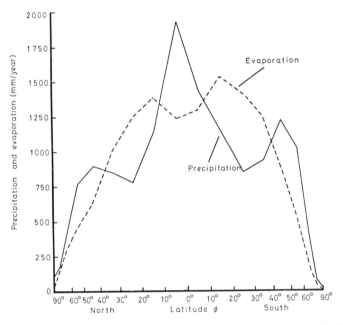

Fig. 10.8. Distribution of mean evaporation and precipitation over the Earth's surface (from data by Sellers).

10.3.3. *Cellular models*

(1) The Hadley cell

The initial difficulty which is met in any attempt to find a coherent explanation of the general circulation, in which cause and effect are so intermingled, is to know where best to break into the system. Hadley argued on the basis of a heat source located at the equator and a cold sink located at either pole. The isobaric surfaces are more widely spaced in the warm air near the equator than in the cold air near the poles; the pressure gradient aloft is therefore directed polewards and the air moves in this direction. The air cools and descends near either pole, and it then moves equatorwards near the surface; it is warmed during this passage and rises at the equator, so completing the circulation. The air is deflected by the Earth's rotation, to the right in the northern hemi-

198

sphere, and to the left in the southern hemisphere: the surface wind is therefore mainly easterly but with an equatorward component, and the upper wind is mainly westerly and has a poleward component.

The resulting circulation is known as the Hadley cell. It cannot represent a steady-state general circulation because easterly surface winds everywhere would imply a continuous slowing down of the Earth's rotation. However, it is in good agreement with what is observed within about 30° of the equator in both hemispheres.

(2) The three-cell model

A difficulty encountered in starting from an assumed heat source at the equator is that the highest mean temperatures are found to occur, not at the equator itself, but at some distance from it. While this is no doubt a 'distortion' produced by the circulation, there is good reason for the view that the feature of the distribution of solar radiation received at the Earth's surface which is most significant is the large horizontal gradient that exists in middle latitudes, compared to both lower and higher latitudes. Strong thermal winds result in middle latitudes to such an extent that the system is dynamically unstable and gives rise to a continual stream of depressions and (fewer) anticyclones†. These pressure features follow many different tracks and have a wide range of life cycles and intensities. There is, however, a strong average tendency for depressions, having first moved mainly towards the east, later to turn polewards then stagnate and fill up in about latitude 60°. The anticyclones or strong ridges of high pressure usually move east, then equatorwards, and merge with predecessors located in about latitude 30°. These events are dictated by the dynamics of the systems and by the scale limitations imposed by geography.

If low average pressure at about 60° and high average pressure at about 30° are accepted as natural consequences of the strong gradients of solar radiation received in middle latitudes, then the other features of fig. 10.2 follow from continuity, even though none may be strictly or obviously imposed by the radiation field. Thus low pressure is required near the equator in order to define the high pressure belts of the sub-tropics; and high pressure is implied near each polar cap.

The winds may be regarded as being determined by this pressure distribution (though pressure and wind are so closely controlled one by the other that the converse may also be argued). Thus within the tropics, easterly surface winds with a component towards the equator comprise the observed trade winds. They converge near the equator where forced ascent carries the air in numerous individual cumulonimbus towers to the tropopause. The air cannot accumulate there and therefore flows back polewards in both hemispheres. As it does so, it is deflected by the Coriolis force into becoming a westerly wind in both hemispheres. The air, displaced polewards in this way and

† This is the third reason referred to in the footnote on page 184.

cooling by net radiation, subsides in the sub-tropical anticyclone; the evidence from tracers is that some also flows on to much higher latitudes. The air which is cooled near the poles sinks and moves equatorwards. It meets warm air, which has diverged polewards from the sub-tropical anticyclone, along the polar front. The interaction of these air masses provides the means by which the warm air ascends to high levels and reaches high latitudes and the cold air penetrates to low latitudes.

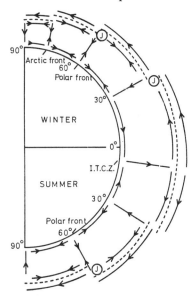

Fig. 10.9. Meridional and vertical components in global motion. The pecked line represents the tropopause; J represents the polar front and winter sub-tropical jets. Horizontal and vertical mixing predominate in the troposphere, associated with the polar fronts and the winter arctic front.

Polewards of about 30°, the mixing effected by cyclones and anticyclones accounts for nearly all the interchange of air that occurs. Within the tropics, cyclones play a much smaller role and the interchange is mainly by systematic meridional motion (though small-scale turbulence plays a vital part near the equator itself). The three-cell circulation shown in fig. 10.9 indicates the variability of direction of transfer (up or down, and polewards or equatorwards) between about 40° and 70°. In the winter hemisphere, air which has descended behind the polar front may move polewards and interact with air, moving from the pole, at the arctic front; this motion may be regarded as comprising a fourth cell.

The angular momentum budget may finally be considered in relation to fig. 10.9. First, we note that the angular momentum of a mass m, moving with angular velocity ω about an axis and at a distance r from it, is given by the product $m\omega r$ and that westerly winds do not increase with latitude at anything like the rate that would be required to keep the absolute angular momentum of the air (i.e. that due to the rotation of the Earth plus that of the air relative to the Earth) from decreasing with latitude. There is therefore a permanent down-gradient of angular momentum directed from low towards high latitudes. It follows that the effect of the mixing of air by synoptic-scale eddies above latitude 30° is to transfer angular momentum polewards, from high to low concentration of momentum; the circumpolar westerly vortex is maintained in this way. For the same reason, mixing in the vertical causes a transfer of this westerly momentum downwards to the ground where it is dissipated by friction. This loss by the air is balanced by the momentum that is imparted to it by its contact with the Earth in the low latitude easterly belt, and the cycle is completed when this momentum is carried upwards and then polewards in the meridional cell, so maintaining the sub-tropical jet.

10.4. *Climatic zones*

We end this chapter with a brief description of the distribution of world climate—a distribution determined by the atmosphere's general circulation, discussed in the preceding sections.

To a first approximation the distribution into climatic zones is based on latitude and is easily explained in terms of fig. 10.2. The zones are as follows.

(*a*) The ' doldrums ' region of light winds, within about 10° of the equator. High temperature, abundant moisture and surface convergence of winds make this a region of copious rainfall of a mainly showery type.

(*b*) The trade wind belts from about 30°–35° equatorwards to the doldrums region. In these zones, fresh winds blow from the north-east in the northern hemisphere and from the south-east in the southern hemisphere. In their poleward sections, more especially in the eastern half of the tropical oceans, the winds are markedly steady and the weather is fine (characterized by flat ' trade wind cumulus '), the easterly current being overlain at a height 1 or 2 km by dry westerly ' anti-trades '. Nearer the equator, and in the more western parts of the oceans, the easterly current is deeper and moister; it is to a greater extent affected by local convergence of winds associated with showers and, mainly in late summer, by tropical cyclones.

(*c*) The sub-tropical high pressure belts of light and variable winds (' horse latitudes '), characterized by descending motion and arid or semi-arid conditions.

(*d*) The belts of ' westerlies ' from about 35°–40° to 60°, south-westerly in the northern hemisphere and north-westerly in the southern hemisphere, but very variable in speed and direction because of the growth and decay of travelling disturbances. The westerlies are, on average, appreciably stronger in the southern hemisphere than in the northern hemisphere—hence the term ' Roaring Forties ' applied to the latitude belt 40° S–50° S. The region is characterized by strong horizontal temperature gradients, in particular by the polar front. The

201

poleward motion of warm air causes widespread and steady precipitation while the equatorward motion of cold air causes showery precipitation.

(e) Polar belts of mainly easterly winds which are by no means free from disturbances. The low temperatures and a predominance of descending motion make this an arid or semi-arid zone.

The important factors which have not been accounted for are the land and sea distribution, topography, and seasonal changes. The resulting distortions of the zones are much greater in the northern hemisphere where the ratio of sea to land is about 60 to 40, than in the southern hemisphere where the ratio is about 80 to 20. The very different thermal properties of land and sea cause particularly large seasonal changes in the northern hemisphere. For example, the summer 'monsoon low' over the Asian continent (centred over N.W. India), caused by the strong differential heating of the land in that season, is replaced in winter by a large anticyclone (centred over Siberia) which is caused by prolonged radiative cooling of the surface. Figures 10.3 and 10.4 show that there is a consequent reversal of flow over S.E. Asia from one season to the other, and these figures also show similar but less conspicuous examples of the same effect in other regions (e.g. N. America).

In middle and high latitudes the effect of the land–sea distribution is to create a natural division of climate which modifies considerably the simple zonal boundaries. Thus a maritime climate is one of fairly high rainfall† and small diurnal and seasonal ranges of temperature and occurs over oceans and western parts of continents; while a continental climate is found over central and eastern parts of continents and has low rainfall and large diurnal and seasonal temperature ranges.

Seasonal changes are important everywhere. The obvious seasonal changes of the idealized general circulation are the shifts of the inter-tropical convergence zone and of the sub-tropical anticyclones 'with the sun', and the broadening and strengthening of the surface westerlies and upper polar vortex in winter. The seasonal changes of a Mediterranean-type climate may be quoted as an example of the last. Winter is the wet season, when the region is affected at intervals by depressions which move from west to east; in summer the weather is dominated by the sub-tropical anticyclone and is very warm and dry.

Even a broad classification of climates over the Earth involves many more sub-divisions than are indicated above. However, to an increasing extent in recent years, climates have not been considered to be adequately defined by such divisions or by simple mean and extreme values of the various weather elements, but have been regarded rather as a synthesis of the day-to-day weather variations over a restricted region in terms of its air mass frequencies and characteristics, including those in the upper atmosphere.

† In these latitudes high rainfall is mainly topographical: rainfall amounts are smaller over the sea although they are, in fact, not well known there owing to difficulties of measurement.

CHAPTER 11

weather forecasting

11.1. *Historical survey*

IN the brief review that follows, the time during which official weather forecasting services have operated is divided into consecutive periods of various lengths, on the basis of their differing main characteristics. These divisions are somewhat arbitrary since there is some considerable overlap between features of adjacent periods.

11.1.1. *1860–1920*

Although local forecasting based on visual observation of the elements is doubtless nearly as old as civilization itself, it was the invention of the telegraph in the mid-19th century which first made organized forecasting possible. The issue of storm warnings and Press forecasts was begun in Britain in 1861.

The weather reports first transmitted by telegraph were very simple, and though pressure was among the reported elements, it appears that some few years elapsed before isobars were drawn on current synoptic charts—obvious though this step now appears. A general association between weather and pressure distribution was then evident and attention became focused on tracking the various types of pressure system as an essential first step in forecasting (as remains true to-day). If, however, as has been suggested, the hope was ever entertained that all that was required to predict the arrival of a depression, with its attendant weather, in the British Isles was to calculate on the basis of its measured speed and direction of motion as it left North America, this was a hope obviously doomed to early disappointment. Disillusionment was, in fact, not long in arriving and forecasts were discontinued for a time; they were, however, restarted (in full in 1879) following a period of detailed study of the connections between the movement and development of weather systems and the associated weather.

Although this period was by no means infertile so far as fundamental ideas in various branches of meteorology were concerned, neither the methods nor the organization for forecasting changed greatly. However, in the last ten years or so, interest in and knowledge of conditions in the upper atmosphere increased considerably under the stimulus of requirements for aviation; also, wirelessed reports from ships at sea were becoming available.

11.1.2. *1920–1945*

Forecasting was, to a very large extent, revolutionized by the techniques of air mass and frontal analysis which were developed in Norway during and immediately after World War I. The ideas coordinated and clarified much that had previously been rather vague, and they were successfully applied throughout middle and higher latitudes. Equally, they were found not to be relevant to tropical meteorology, although a special kind of front may be considered to occur in the inter-tropical convergence zone in which the trade winds of the two hemispheres meet.

During this period there was increasing use of upper-air temperature, humidity, and pressure measurements. Prior to 1940 these were made in special aircraft ascents and were limited both in their number and in the height which they reached. The subsequent introduction of the radiosonde provided more detailed information and, although the war-time ban on international exchange of meteorological information restricted the synoptic use of these observations, they were invaluable in the tephigram type of analysis described in Chapter 5.

11.1.3. *1945–1960*

The extension of radiosonde and radar-wind observations early in this period (including those from Ocean Weather Ships) allowed synoptic upper-air analysis to be made for a large area. The movement and development of systems, which had previously been judged almost entirely on the basis of experience with the surface synoptic chart, were now related to patterns of thickness contours and absolute contours (Chapters 6 and 8). The significance of the polar front jet stream, the existence of which was apparently unsuspected† prior to some war-time incidents which involved military aircraft in N.W. Europe and Japan, was now realized, and the phenomenon was intensively studied because of its importance to aviation and in synoptic development.

Forecast charts were prepared for the surface and selected upper levels for periods 24 and 48 hours ahead; the expected positions of isobars and fronts were shown on the surface chart and those of contours on the upper-air charts. These charts—especially the surface chart—were the basis of the forecast issued for 24 hours ahead and of the ' further outlook ' (usually for a further day). If they were significantly in error for

† (a) This may appear surprising in view of the frequent pilot balloon observations that were made prior to 1940. However, in lightly-clouded skies, observation ceases when the balloon is lost in the distance; this means that, in strong winds, the balloon cannot be followed to jet-stream levels. A similar, though less serious, bias occurs with radar tracking of balloons, so that there is even now an acknowledged danger of underestimation of the frequency of strong upper winds.

(b) Again, nephoscope observations (involving measurement of the angular velocity of clouds as mentioned in §4.3.3) were frequent prior to 1940. Presumably, the very high speeds measured for some examples of jet-stream cirrus were discounted because of the more uncertain nature of these measurements and the narrow belt of high speeds involved.

a particular area then the forecast weather for that area would almost certainly be wrong.

In the latter part of World War II, ground radar had already been used, in some regions with few observing stations, to detect convective storms and to display their positions. (Radar echoes from the larger drops associated with moderate or heavy rain may be obtained out to a distance of about 100 miles from the radar location: an aerial system which scans horizontally is used and the position of echoes obtained over this region is presented, in plan view, on a cathode ray screen.) During the period under discussion, weather radar was installed at some airfields and large centres of population. With its aid, the forecaster was able to observe the movement and development of rain belts in detail and to make short-period local forecasts with a precision and degree of confidence which would otherwise have been impossible, especially in showery and thundery types of situation. A more recent illustration of a weather radar picture is contained in fig. 11.1.

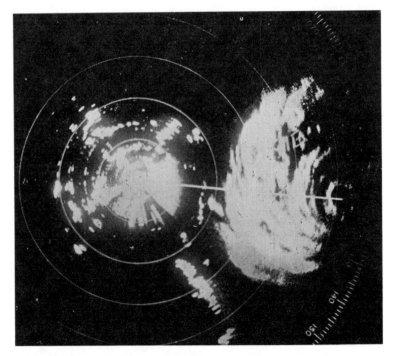

Fig. 11.1. Weather radar photograph of typhoon 'Elsie', Hong Kong Royal Observatory 15 September, 1966. The radar is installed inland at a height of nearly 600 metres and has a horizon of about 200 miles; the range markers are 40 n miles apart. The eye of the typhoon is near the radar horizon and the spiral rain bands are clearly shown.

11.1.4. *1960 onwards*

The period since about 1960 has witnessed a revolution in the method used to obtain the forecast charts referred to in the last section. The immediate stimulus was provided by the advent of electronic computers, but the method itself stemmed from work done much earlier, during a period of more than 10 years, by an Englishman, L. F. Richardson.

Richardson was, in a sense, a generation ahead of his time. The end which he foresaw was the prediction of the surface pressure field by the application of basic physical laws, in numerical form, to an initial observed state of the atmosphere. He attempted to justify these ideas by predicting the 6-hour surface pressure change, averaged over an area of western Germany, from an initial distribution of pressure in N.W. Europe. This test, limited though it was, involved him in many thousands of calculations, and the pressure change which he finally obtained was quite different from that which actually occurred (a predicted rise of about 60 mb in 6 hours compared to a very small actual change).

The theory and practice of the method were described in his book, 'Weather Prediction by Numerical Process', published in 1922. Richardson also discussed there the reasons for the failure of his forecast; inadequacy of initial observations was one reason but other factors were also involved. Nevertheless, he envisaged a time when a small army of human computers would be employed in continuous calculations to keep ahead of world weather.

When electronic computers became available in the 1950s, meteorologists began to reconsider the prospects of obtaining numerical predictions. The initial experimental forecasts in that period, made in the U.S.A., were for a single level in the middle troposphere and were sufficiently encouraging for research into the development of atmospheric models suitable for numerical treatment to be taken up in various countries.

In the British Meteorological Office, methods of numerical prediction were tried out for a number of years during which the established methods of chart prediction were continued as the forecaster's main guide. Comparison of the respective forecasts with actual contours and isobars showed that the numerical method was, on average, the better of the two for the upper atmosphere and that there was little to choose between them for the surface. In the latter part of 1965 the numerical method was adopted for both surface and upper levels as a routine†. The change is no doubt a permanent one because the subjective method, dependent as it is on individual skill and experience, has probably reached about its limit of reliability, whereas the numerical method is relatively new and is capable of being developed further.

† A forecast of the distribution of mean temperature (layer thickness) is also implied when the pressure distribution is forecast at two levels (see section 6.5.3).

206

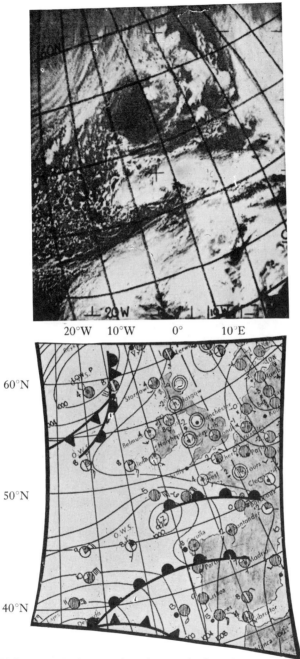

20°W   10°W   0°   10°E

60°N

50°N

40°N

Fig. 11.2.   A.P.T. photograph and synoptic chart for 17 February, 1969.
(Photograph is Crown copyright, reproduced by permission of the Controller of H.M.S.O.)

207

The human forecaster has, however, by no means been eliminated. Accurate forecast charts are a more-or-less necessary, but by no means sufficient, condition for obtaining accurate predictions of weather elements. The forecaster's skill is still required to interpret the charts.

In the 1960s, benefit has been derived from various technical advances, outstanding among them being the remarkable American satellite cloud photographs which may be intercepted daily in all parts of the world. They illustrate many features of cloud structure and formation; while, in regions of few ground observations, they are especially valuable in the information they give on convective activity, the positions of fronts and depressions, and the extent of stratus or fog, snow or ice. An example of a photograph obtained by the current A.P.T. (Automatic Picture Transmission) system, on 17 February 1969, is shown in fig. 11.2, with the corresponding synoptic chart for the same area (on a slightly different projection). There is extensive snow cover over the British Isles, where cloud amounts are small. The close relationship between the surface chart (pressure systems and fronts) and the cloud distribution is striking, while the direction of the circulation is clearly revealed by the spiralling bands of clouds.

11.2. *Conventional forecasting*

The method of forecasting based on the somewhat subjective judgment of synoptic charts, and of diagrams such as the tephigram, is here termed conventional in order to distinguish it from objective methods which have been more recently adopted to some extent.

11.2.1. *Pressure tendency*

In synoptic meteorology, surface isobars and upper-air contours are predicted by extrapolation from the recent history of the charts concerned, and by inferences made directly from current charts on the lines indicated in chapter 8. In practice, a good deal of weight is attached to the pressure tendencies (i.e. 3-hour changes) which are plotted on the surface chart.

An observed pressure tendency may, in general, be regarded as having separate contributions from the component of velocity of a pressure system towards the station concerned and from the simultaneous development of the system†. To quote an example: a deepening low causes a bigger pressure fall at a station in its path than does a low, with similarly distributed isobars and the same velocity, which is not deepening. An equation may be given which expresses this fact. It is analogous to that given in problem 8.4, with the pressure system taking the

---

† In a ship report of tendency, an allowance is usually required for movement of the ship across the isobars in the 3-hour period concerned.

place of an individual element of air. In finite-difference form the equation is

$$\frac{\Delta p}{\Delta t} = \frac{\delta p}{\delta t} - c\,\frac{\Delta p}{\Delta s} \qquad (11.1)$$

where $\Delta p/\Delta t$ is the observed local pressure tendency ($\Delta t = 3$ hours), $\delta p/\delta t$ is the change of central pressure of the system in the same time interval (i.e. its rate of deepening or filling), c is the speed of the system, and $\Delta p/\Delta s$ is the pressure gradient (rate of pressure rise with distance) in the direction of c.

A forecaster attempts to distinguish, though not usually in strict quantitative terms, the movement and development parts of the tendency field. In an unsettled westerly situation the effect of movement dominates, while the reverse is often true in more static situations. A sustained and general rise or fall of pressure over a region is nearly always significant—especially so because it often indicates a change of synoptic type. Thus, such a fall in the central regions of an anticyclone usually indicates its imminent collapse and a change to rainy and, in summer, thundery weather. On the other hand, a general rise of pressure in the warm sector of a depression is a strong indication that a warm anticyclone will soon be established there. The significance of a sustained surface pressure tendency is that it implies a specific distribution of divergence and a particular sign of vertical motion in the troposphere—upwards with a pressure fall, downwards with a pressure rise (see § 8.7.1)†.

Pressure tendencies are complicated to some extent by atmospheric motion of a tidal nature. The atmosphere, being fluid, responds to the gravitational pull of the moon (and sun) as the oceans do. The air is heaped up, and pressure is therefore relatively high, 'under' the moon and also in the opposite longitude, and it is low in the longitudes 90° away. When the pressure values are arranged according to their lunar hour, and are then averaged, the range of pressure within the lunar day seems to be far too small (only about 0·2 mb in low latitudes and less elsewhere) to be of any practical forecasting significance‡.

Much larger regular oscillations of the atmosphere occur within the solar day. As with the oceans, the effect of the gravitational pull of the sun is less than that of the moon. However, unlike the oceans, the atmosphere warms and cools, and therefore also expands and contracts, and converges and diverges, as it rotates under the sun. The corresponding surface pressure variations reach maxima at about 10 and

† In the rear of a depression, showeriness initially increases with rising pressure because the depth of cold unstable air increases. The large-scale subsidence of air which accompanies a sustained pressure rise causes the showeriness subsequently to die out.

‡ Although this is true, a small statistical influence of the moon's phases on the incidence of rainfall seems recently to have been demonstrated for many parts of the world.

22 hours local time and reach minima at about 04 and 16 hours. These variations are conspicuous on the barograph charts in low latitudes (their daily range is a few millibars there) and they complicate the use of the pressure tendency in tropical forecasting: tendencies are considered there either in comparison with the normal changes for the 3-hour period concerned or are calculated for a period of 24 hours. Complications are much less serious in middle and higher latitudes because the tidal pressure fluctuations are smaller and because the general synoptic ' noise level ' is greater. In a region which lacks, for a time, both synoptic movement and development, the diurnal pressure variation produces small tendencies of opposite sign on successive 6-hour charts.

11.2.2. *Making the forecast*
(1) *General*
When the forecaster has drawn his anticipated surface chart for a period 24 hours ahead, say, how does he decide what the associated weather will be?

He thinks mainly in terms of air masses and (if applicable) of fronts. If, as is usual, he has current information about their characteristics he assumes as a first approximation that these will be translated to the region with which he is concerned; however, he also has to assess and allow for the various modifying influences which will operate, these being of a physical or dynamical or local or diurnal nature.

The air may be heated or cooled, moistened or dried, by surface contact. There may be anticipated growth from a small wave disturbance to a large disturbance (in which case the success of the prediction of the elements is very dependent on accuracy of the forecast chart). The forecaster may anticipate large-scale vertical motion of the air which may greatly modify the stability and humidity characteristics currently observed in the air mass. He must also take into account the different degrees of exposure and shelter, and of forced orographic ascent and descent, which will operate in different areas in the anticipated wind conditions. He has to allow for the convective build-up in polar masses overland during the day and their dispersal at night; also for the possibility that outward radiation may then produce fog or frost: conversely, he must allow for day-time dispersal of stratus cloud by surface heating and its reappearance at night.

(2) *Unsettled westerlies*
Figure 11.3 illustrates a typical sequence of westerly-type weather over the British Isles. The period concerned was the 5 days from 12 to 16 November, 1966 during which various frontal troughs and weak ridges crossed the country. The synoptic chart sequence is shown at the bottom of the diagram along with a map which indicates the positions

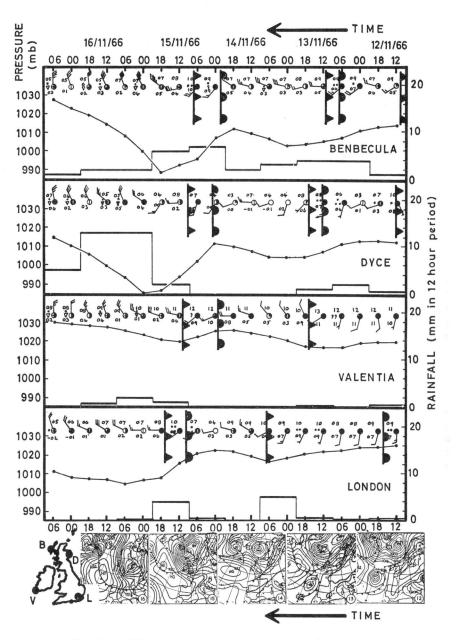

Fig. 11.3.   Time-section of weather elements at four stations
12–16 November, 1966.

211

of four stations (B, D, V, L) selected as representative of different parts of the country during this period.

The sequences of shortened synoptic plots (weather, wind, cloud amount, temperature and dew point—compare fig. 8.21) are shown for each station. Note that the time sequences of the weather elements and of the synoptic charts read from right to left: it is, therefore, as if a station were transferred from right to left through a pattern of ridges and troughs which has been suitably modified for the particular stage of development of the system, and time of day, at which the station moves through it. The times of frontal passage are indicated. Rainfall amounts in 12-hour periods are shown by a histogram, and a graph of pressure variation with time is plotted—this being the mirror image of the barograph trace at the station concerned.

On 12 and 13 November the lows in the Iceland–Greenland region moved eastwards then turned off left, filling up slightly†. The new low (P), which appears south-west of Greenland on the 13th, deepened and moved in a direction north of east, over about 40° of longitude, in the next 2 days to reach a position just south-east of Iceland on the 15th. It then changed direction abruptly and moved south-eastwards to the coast of N.W. Germany within the next 24 hours, its central pressure meanwhile rising by 12 mb. (This heralded a change of weather type, and a ridge of high pressure subsequently extended from Norway to the British Isles.)

Attention is drawn to the following features of the plotted observations and graphs.

(a) Pressure is, on average, 15–20 mb higher at the two southern stations (V and L) than at those in the north (B and D). This is reflected in the rainfall amounts for the 5-day period: Valentia 4·5 mm, London 9·9 mm, Benbecula 32·0 mm and Dyce 36·1 mm.

(b) The effects of exposure and shelter are seen in the larger amounts of rain at Benbecula, compared to Dyce, when winds are south-west to west, and in the converse with northerly winds: (the effect of the latter is accentuated at Dyce by the proximity of the low).

(c) Rainfall tends to occur mainly at times of pressure fall; however, the largest amounts of rainfall occurred at Dyce in the heavy showers, accentuated by topography, in the arctic air mass behind the final trough.

(d) The main forecasting problem in a westerly situation such as this is one of accurate timing of troughs and ridges. However, in the case under discussion, the sudden change of direction of the low on the 15th would have had to be foreseen if the distribution of showers on the 16th, and the subsequent change of synoptic type, were to be successfully predicted.

† The depression which lies north-west of Spain on the 12th and disappears by the 14th is of interest in that it is an old tropical cyclone.

(3) *Forecasting problems*

Mobile synoptic situations such as that just illustrated are obviously calculated to keep the forecaster well occupied. In more settled situations his problems are less obvious but no less real. Radiation fog and frost, for example, are common in anticyclones in the British Isles during at least the winter half-year; while in summer, an anticyclone has barely to be established before the risk of the occurrence of coastal fog (by advection) in one region or another has to be considered.

The forecaster's greatest challenge lies in the prediction of such events as rainstorms of flood proportions, immobilizing snowstorms, or violent gales: moreover, these are the occasions when his reputation is at greatest risk. Such exceptional events occur when there is an unusual combination of 'favourable' circumstances. The forecaster's difficulty is to decide, almost entirely subjectively and in a limited time, just how exceptional a particular set of circumstances is. It is difficult to see how, in the present state of knowledge, these events can be regularly predicted without also over-warning in less exceptional cases.

The forecaster is criticized for the vague terms in which his predictions are sometimes expressed. He may well reply, in his own defence, that generality, and therefore also vagueness, are forced on him by the fact that he has a very limited time in which to describe variable conditions over a large region. If pressed, he would probably admit, however, that there are occasions when he is relieved that time does not permit of his being specific as to whether there will be fog or 'no fog', or rain as opposed to snow, or thunderstorm or 'fine', in a particular locality. These are limitations which are imposed on the forecaster by the very nature of the weather, as is easily understood by even casual observation of these phenomena on many occasions. It is almost certain that in marginal conditions nothing better than shortest-period prediction will be possible at any time in the future; even this much will depend on the development of further aids such as weather radar.

When the recipient of a forecast has a sound background knowledge of meteorology, he is able to recognize those types of synoptic situation in which prediction is inherently more difficult and uncertain than normal. If he couples this knowledge with observation of the local elements, it is usually possible for him to assess the situation intelligently and somewhat in advance of events.

11.3. *Long-range forecasting*

In general, it is not possible to be at all definite about the weather over the British Isles (or for a comparable mid-latitude region) for more than one or two days ahead. When, however, the synoptic situation is dominated by an anticyclone which is centred over or near the British Isles, a confident forecast is often possible for three days ahead for the country as a whole, and occasionally for four or five days for some parts of it. This represents the limit of prediction by conventional methods.

What, however, is the prospect of obtaining reliable forecasts—even if expressed in quite general terms—for longer periods by other means?

Apart from the inherent interest of this problem, the great economic and other benefits that would result from its solution have ensured that it has received much attention over many years and in many parts of the world. An account of this work is inevitably monotonous in the (almost) uniform lack of success that has attended it. Nevertheless, we review the work briefly in order to indicate the various ways in which a solution to the problem has been sought.

11.3.1. *Statistical methods*

An obvious first step is to ' let the atmosphere speak for itself ' by statistical analysis of accumulated weather observations. There have been many attempts to do this, especially in relation to the surface elements of pressure, temperature and rainfall.

(1) *Periodicities*

Sir Gilbert Walker, who was among the most assiduous of researchers into meteorological statistics, wrote that faith in the existence of periodicity in the weather (quite apart from the seasonal changes of the elements) is instinctive in man. It is a faith which is, however, difficult to retain in the light of the unsuccessful searches that have been made for significant periodicity, the periods ranging in length from a few days to many years. There is a distinct possibility that any periodicity that exists is not persistent but rather may last for a time, then disappear, then reappear after an interval but with a changed phase. There have, in fact, been suggestions that something of this kind happens in middle latitudes, the period concerned being between 20 and 30 days. However, even if real, this tendency is not strong enough to be of practical significance in forecasting. The same appears to be true of any other period that has been suggested.

(2) *Solar connections*

As was mentioned in Chapter 3, the sun's activity fluctuates in an obvious and more-or-less regular way, with a period that averages about 11 years. The reason for this periodicity is unknown but it demonstrates that the idea of an underlying rhythm within the earth's atmosphere has to be considered seriously, even apart from the obvious possibility of a direct influence of solar variability on the elements. There is no doubt that variations of this latter kind occur in the highest parts of the atmosphere but it is much more doubtful if there is any measurable effect within the troposphere or lower stratosphere. The position in the latter respect is rather controversial. Some forecasting methods have, in fact, been based on supposed solar connections with the weather but appear to have made no impact.

(3) *Climatic singularities*

Features of local climate that occur around the same calendar dates in most, if not all, years are called singularities. If they are reliable, then they should obviously be taken into account in the framing of forecasts. However, in middle latitudes at least, no singularities are completely reliable, and even where they are real they tend to be rather variable in their time of occurrence. However, they have some limited value in forecasting because, for a period well ahead at least, they (being part of the normal climate of the region concerned) are probably the best estimate that can be made of the future weather in such circumstances.

(4) *Contingencies*

A question that arises naturally on most occasions of a spell of weather of a particular kind is the significance such a spell may have in relation to the month or season ahead: thus, does a very cold November make it more likely or less likely that the following December, or the following winter season, will also be colder than normal?

This type of question is easily answered, in terms of statistical probability, by an examination of past records. It may be, for example, that out of the 10 known very cold Novembers at the place concerned, 6 were followed by colder-than-average (and 4 by warmer-than-average) Decembers. The probability of a colder-than-average December may then be given as 60%. It is obvious that this approach is limited in various ways. No reference is made to any element other than temperature or to conditions in any other region, either of which might have given better discrimination. The approach is even less sophisticated than the analogue method to be mentioned later; it is contained, in principle at least, within that method but is only a small part of it.

This type of study need not be confined to contingencies between events at the same place or in the same region. Walker made some early studies of the correlation between conditions in one region with earlier conditions (often involving quite different elements) in other, perhaps distant, regions. This work showed that there are world-wide connections between meteorological conditions but did not bring to light correlations which were strong enough, or stable enough, to be useful in prediction over a long period.

11.3.2. *Synoptic methods*

Determined efforts have been made in some countries—in Russia and Germany, for example, but especially in the U.S.A.—to adapt the synoptic method of short-period forecasting into one applicable to longer-range forecasting.

Minor pressure systems are eliminated, in the American routine system, by concentrating on average circulation at a height of about 3 km. A 30-day forecast is aimed at, based on the corresponding predicted mean circulation. This latter is arrived at by various methods which include extrapolation, statistics, dynamical reasoning and also the use of successive five-day-mean circulation charts. The end-product is the predicted distribution of temperature and rainfall for 30 days ahead over a particular region, this being determined from past statistics which relate these elements to mean circulation for the region.

It is clear that uncertainties involved in the method are large, not only in predicting the mean circulation but also in inferring the distribution of temperature and rainfall from the circulation. It seems, nevertheless, that it achieves some degree of over-all success.

11.3.3. *Analogues*

A method of 30-day forecasting which is, in principle, much simpler than the American method, but which is perhaps no less involved to apply in practice, is used in the British Meteorological Office.

The method is based on finding from past records an analogue for the month just past and then forecasting for the coming month the weather which followed the analogue month. More than 80 years' data are used in the search for good analogues which are selected for about the same time of year as the month concerned. Comparisons are made between the distributions of 1000–500 mb thickness and mean-sea-level pressure over most of the northern hemisphere, and also between the synoptic chart sequences over western Europe and the eastern Atlantic. Special importance is attached to the fact that the development which occurred within the first few days that followed a selected analogue agrees with what is expected for the month under consideration.

The method is streamlined by the use of electronic computers. On some occasions several good analogues may be found; if the weather which followed each of these was similar, then the terms of the forecast are not in doubt and there are good grounds for confidence in it. Conversely, if no good analogues are found and if different patterns of weather followed the best that are available, then there is little or no basis for issuing a forecast and there can be little confidence in that which is issued. Usually, the situation is between these extremes, and the analogue is selected by consensus opinion which may take account of additional factors such as the distributions of sea-surface temperature and of polar ice.

The actual forecast issued is in general terms and includes a statement concerning the character of the month as a whole and the expected temperature and rainfall, sometimes also the sunshine, in different regions, relative to the normals for the month†. The success of the

† The ' Weather Log ' contained in the Royal Meteorological Society monthly magazine ' Weather ' reproduces the official forecasts along with background information, including monthly mean charts of surface pressure and of 1000–500 mb mean temperature, for the northern hemisphere. (The log includes a monthly summary of selected elements for the British Isles and also daily weather maps).

forecasts is difficult to assess but is certainly higher for temperature than for rainfall, the latter being notoriously difficult to pinpoint accurately. An overall success rate of two forecasts substantially correct out of three has been quoted.

11.4. *Numerical forecasting*

Numerical forecasting involves concepts in mathematical physics and numerical analysis, many of which are outwith the scope of this book. Since, however, the method is already important and will become increasingly so, some explanation of its principles and practice is required.

Numerical prediction starts from an initial observed state—usually the contours of one or more isobaric surfaces—over a large area. This being for the hour $t=0$, then the corresponding state 24 hours later $(t=24)$, say, is predicted without reference to conditions prior to $t=0$; thus, in contrast to the conventional method, 'history' plays no part.

The prediction is arrived at by setting up and solving a forecasting equation, the form and complexity of which depend on the particular 'model' of atmosphere assumed. The equation concerned gives the local rate of change of contour-height, i.e. $\partial Z/\partial t$, in terms of the observed horizontal gradients of Z, i.e. $\partial Z/\partial x$ and $\partial Z/\partial y$. However, there is no exact solution of an equation of this kind, the underlying physical reason being the cross-linked nature of atmospheric processes; hence conditions at $t=24$ cannot be obtained exactly and in a single step from $t=0$. An approximate, step-by-step method of solution is therefore used. The principle and method may be illustrated by describing the barotropic model by means of which useful mathematical prediction was first shown to be possible.

11.4.1. *The barotropic model*

A barotropic atmosphere is one in which there are no thermal winds; this model therefore takes no account of the type of development mentioned in §8.7 and refers in fact, to a single level (usually 500 mb). It has other restrictions: (a) the movement of the air after $t=0$ is governed solely by the inertia it possesses at $t=0$, i.e. no allowance is made for fresh energy input; (b) the motion is assumed to be entirely horizontal; (c) the motion is assumed non-divergent, so that each particle conserves its initial absolute vorticity during its motion (see §8.7.3); (d) the vorticity is measured, not from the actual winds but from the geostrophic winds which are derived from contours; this 'geostrophic vorticity' (ζ_G) is also assumed to be transported by the geostrophic winds (V_G) and not by the actual winds.

To determine the corresponding forecasting equation consider first ζ_G. We have, from a comparison with equation (8.12)

$$\zeta_G = \frac{\partial v_G}{\partial x} - \frac{\partial u_G}{\partial y}. \tag{11.1}$$

Now, the components of V_G are

$$u_G = -\frac{g}{f}\frac{\partial Z}{\partial y} \quad \text{and} \quad v_G = \frac{g}{f}\frac{\partial Z}{\partial x}, \qquad (11.2)\dagger$$

$$\therefore \zeta_G = \frac{\partial}{\partial x}\left(\frac{g}{f}\frac{\partial Z}{\partial x}\right) + \frac{\partial}{\partial y}\left(\frac{g}{f}\frac{\partial Z}{\partial y}\right)$$

from equations (11.1) and (11.2);

$$= \frac{g}{f}\left(\frac{\partial^2 Z}{\partial x^2} + \frac{\partial^2 Z}{\partial y^2}\right)\ddagger$$

i.e. $\zeta_G = \frac{g}{f}\nabla^2 Z,$ \qquad (11.3)

where

$$\nabla^2 = \frac{\partial^2}{\partial x^2} + \frac{\partial^2}{\partial y^2}.$$

Considering now absolute vorticity, and noting that the partial differential signifies local change and that $\partial f/\partial t$ is therefore zero, we have

$$\zeta_G + f = \frac{g}{f}\nabla^2 Z + f$$

$$\therefore \frac{\partial}{\partial t}(\zeta_G + f) = \frac{g}{f}\nabla^2\left(\frac{\partial Z}{\partial t}\right). \qquad (11.4)$$

The change of absolute vorticity following the motion is related to its local change and to that produced by advection by the formula that was given (for temperature) in problem 8.4; thus

$$\frac{\mathrm{d}}{\mathrm{d}t}(\zeta + f) = \frac{\partial}{\partial t}(\zeta + f) + V.\frac{\partial(\zeta + f)}{\partial s}. \qquad (11.5)$$

For geostrophic absolute vorticity which is advected by the components of geostrophic motion, this becomes

$$\frac{\mathrm{d}}{\mathrm{d}t}(\zeta_G + f) = \frac{\partial}{\partial t}(\zeta_G + f) + u_G\frac{\partial(\zeta_G + f)}{\partial x} + v_G\frac{\partial(\zeta_G + f)}{\partial y}. \qquad (11.6)$$

Now, because the flow is assumed non-divergent,

$$\frac{\mathrm{d}}{\mathrm{d}t}(\zeta_G + f) = 0$$

from equation (8.15);

† From equation (6.4): verify that these components have the correct signs; e.g. when $\partial Z/\partial y$ is positive (i.e. high values of contour to the north) in the northern hemisphere (where f is positive) then u_G is negative, i.e. the component is from the east and keeps low contour values on the left.

‡ This step assumes f constant and therefore neglects a small term which arises because of the variation of f in the y direction.

$$\therefore \frac{\partial}{\partial t}(\zeta_G + f) = -\left[u_G \frac{\partial(\zeta_G + f)}{\partial x} + v_G \frac{\partial(\zeta_G + f)}{\partial y}\right]$$

from equation (11.6);

$$\therefore \frac{g}{f}\nabla^2\left(\frac{\partial Z}{\partial t}\right) = -\frac{g}{f}\left[\frac{\partial Z}{\partial x} \cdot \frac{\partial(\zeta_G + f)}{\partial y} - \frac{\partial Z}{\partial y} \cdot \frac{\partial(\zeta_G + f)}{\partial x}\right]$$

from equations (11.2) and (11.4);

i.e.
$$\nabla^2\left(\frac{\partial Z}{\partial t}\right) = \frac{\partial Z}{\partial y} \cdot \frac{\partial(\zeta_G + f)}{\partial x} - \frac{\partial Z}{\partial x} \cdot \frac{\partial(\zeta_G + f)}{\partial y}, \tag{11.7}$$

where

$$\zeta_G = \frac{g}{f}\nabla^2 Z.$$

Equation (11.7) is the forecasting equation for the barotropic model. All the quantities on the right-hand side of the equation can be evaluated, at a selected point, from the distribution of Z at and around the point, each partial differential being replaced by a finite difference. The calculations are carried out at each of the points (perhaps 1000) of a uniform grid, each separated by a distance (d) which is of the order of 300 km. The area covered by the calculations is appreciably bigger than that for which prediction is required, because the latter must not be reached, in the forecast period, by unknown conditions which are transported across the boundaries of the grid. Initial values of Z at 500 mb are interpolated, at each grid point, from the radiosonde network observations. Each ζ_G is calculated from equation (11.3),† the appropriate value of f is added, and the right-hand side of equation (11.7) is computed at all grid points (apart from outer rings which are 'lost' in taking finite differences).

These calculations give the distribution of $\nabla^2(\partial Z/\partial t)$. The required values of $\partial Z/\partial t$ at every point are then obtained by a long numerical process of trial and error which is continued until a distribution of values of $\partial Z/\partial t$ is obtained which fits the gradient distribution of tendency, $\nabla^2(\partial Z/\partial t)$, over the whole area. The individual values of ΔZ for a small Δt (e.g. one hour) are found from the values of $\partial Z/\partial t$ and they

† $\nabla^2 Z$ is approximated by

$$\frac{\Delta}{\Delta x}\left(\frac{\Delta Z}{\Delta x}\right) + \frac{\Delta}{\Delta y}\left(\frac{\Delta Z}{\Delta y}\right).$$

The student may verify from this formula that $\nabla^2 Z$ at a central point C, in which equidistant grid points A, B, C, D, E read from west-to-east and equidistant grid points F, G, C, H, J read from south-to-north, is

$$(\nabla^2 Z)_C = \frac{Z_E + Z_J + Z_A + Z_F - 4Z_C}{(2d)^2}$$

d being the distance between adjacent points.

219

are added algebraically to the initial values to give a new distribution of Z for $t = 1$, i.e.

$$Z_1 = Z_0 + \left(\frac{\partial Z}{\partial t}\right)_0 \Delta t.$$

The whole process is repeated again and again with the same time-interval until the end of the forecast period is reached.

It is obvious that a prediction of this kind involves so many separate calculations that it is a possible forecasting method only if an electronic computer is available. In this barotropic model the operation of the machine is entirely in terms of Z. The physical interpretation of the machine procedure is that the original field of motion is changed, step by step, as each particle of air moves, carrying with it its individual rate of absolute spin. Realistic results were obtained with this first (out-dated) model because at the 500 mb level, to which it was applied, flow is often nearly non-divergent.

11.4.2. Later developments

Subsequent work has concentrated on the removal of many of the restrictions of the barotropic model. In particular, various baroclinic models have been used in which conditions at two, three, or more levels are considered: in this way, account is taken of the synoptic developments which are associated with thermal winds. There has also been research into the inclusion of such influences as topography, surface heating, surface friction, and latent heat exchanges; also on the use of the wind field itself rather than on its spin as described above. Finally, there have been advances in technique in order to speed up the whole operation. Thus, for example, the synoptic messages are fed at high speed directly into the computer which is programmed to accept only certain types of information, to detect obvious errors in the data and to interpolate the appropriate values of contour-heights at grid points—all prior to its solving the forecasting equation.

11.5. Predictability and control

The limitations of present-day forecasting are rather well known generally and are freely acknowledged by the meteorologist. On the whole, they are not limitations for which he holds himself responsible—despite the proprietary attitude to the weather that is sometimes adopted, and is no doubt difficult to avoid, by meteorologists who present forecasts to the public. A facile view of science is that all things are possible, if not tomorrow then the day after. This view has been expressed with respect to weather forecasting and control: to what extent is it justified?

11.5.1. Short-range predictability

We earlier expressed the view that forecasting of various marginal events, with a reasonable degree of confidence, will always be impossible for more than a few hours ahead at best. Even this much is unattainable in the case of events which are basically of a statistical nature: thus, there is no ' explanation ', far less possibility of prediction, of an individual gust or shower—as opposed to a general degree of gustiness or showeriness.

What of forecasts for a period about 24 hours ahead? Three separate possibilities of improvement may be considered; namely, more detailed information about the atmosphere's initial state, more accurate numerical prediction of pressure, and more accurate interpretation of the predicted chart. There is no assurance that better information will, in fact, improve 24-hour forecasts to a significant degree: J. S. Sawyer, Director of Research in the British Meteorological Office, for example, writes that ' —many of the present forecast failures are not due to lack of information—our present understanding is insufficient to predict the changes however complete our knowledge of the initial state of the atmosphere '. On the other hand, there are grounds for the belief that the accuracy and dependability of numerical forecasting will improve, though to what extent is not known. Finally, ' filling in ' of the weather by the human forecaster is unlikely to advance significantly; individual forecasters of long experience have in the past set such high standards as are difficult even to maintain. There is a possibility that objective forecasting methods, based on established principles and being a kind of distillation of all forecasting experience, may be derived from an examination of past records. This has been tried, on a very limited scale, with some success—as in forecasting minimum temperature, or the occurrence of fog, at particular sites. It seems likely, however, that extension of this method will be confined, for a long time to come, to certain specific problems which are only a small fraction of all the situations which confront the forecaster.

11.5.2. *Medium-range predictability*

If for no other reason than that the initial state of the atmosphere can never be perfectly observed, perfect forecasts extending indefinitely into the future are an impossibility; even with an exact forecasting equation, the solution for an observed atmosphere must diverge continuously from that of the real atmosphere. There is, however, an even more fundamental time-limitation to predictability, in that the smallest scale of atmospheric motion—which by its very nature can never be observed—ultimately affects the motions of even the largest scale, namely the giant depressions and anticyclones.

So far as the period of useful forecasting is concerned, the vital questions are how quickly the solutions for the observed and real states diverge, and how long it takes the smallest-scale motions to affect those of large scale. Numerical experiments, carried out mainly in the United States under J. Smagorinsky and under E. N. Lorenz, have been interpreted as indicating that useful prediction about day-to-day weather may be possible up to about three weeks ahead but no longer. Research meteorologists are far from being unanimous on this question: some consider the attainable period to be appreciably greater than this, while others take the view that the atmosphere is inherently unstable to a degree that makes such a prediction period unattainable.

By ' useful day-to-day prediction 'it is not implied that detail such as rainfall amount or probable times of beginning and end of rainfall are envisaged, but rather that the main features of the circulation at the surface and in the upper air may be known with some confidence throughout the period concerned, thus allowing of an interpretation of regional weather, and of rainfall and temperature relative to normal. Reliable forecasts of such content would be extremely valuable. R. C. Sutcliffe has made a point which highlights the magnitude of the problem that would be involved in accurate long-period forecasting of rainfall amount from first principles. The average precipitable water in an atmospheric column is about 25 mm and the average annual rainfall is about 900 mm. This means that the water ' turns over ' between atmosphere and Earth about 36 times in a year. Thus accurate prediction of rainfall amounts for one month ahead may perhaps be regarded as akin to tracing the paths of individual air particles and calculating the details of their passage, three times, through the precipitation-evaporation cycle.

11.5.3. *Long-range predictability: climatic trends*

The synoptic and analogue methods of 30-day forecasting described in §11.3 do not appear capable of being further developed to any significant extent. It is scarcely an exaggeration to say that in all other respects the meteorologist has, at present, nothing positive to contribute concerning future longer-period weather prospects†. This situation is very unlikely to change until numerical studies of the general circulation, mentioned in Chapter 10, are much further advanced than now.

The problem of long-range forecasting merges with that of predicting climatic trends. Apart from the intrinsic interest of past climates, it is essential, in considering future trends, to examine the changes that have occurred in the past and, if possible, determine the causes of these changes. There is a wide variety of evidence on past changes; for pre-historic times it is mainly geological and biological in nature— relating, for example, to coal and ice deposits and to the distribution of plants and animals—and indicates short Ice Ages and much longer warm periods without permanent ice. In the historical period, other types of evidence are taken into account—evidence of social and agricultural changes, of tree rings indicating rates of growth in long-established

† There is a small statistical tendency for persistence of average warmth or cold from one month to the next, even from one season to the next, at least in the British Isles. However, there are many exceptions.

specimens, of variations in the levels of seas and salt lakes; finally, in the last 200 years or so, the evidence has included instrumental records and also direct observations of the advance and retreat of glaciers, these last being a most sensitive integrator of climatic fluctuations.

Between about 1850 and 1940 a slow and irregular warming trend is discernible in the records; the temperature rise was bigger in medium and higher latitudes than in low, and averaged no more than 1°C over the Earth as a whole. Modest though such a trend may appear, it has marked effects in polar and sub-polar regions. Such a temperature rise represents a large amount of additional energy when totalled for the whole Earth and it is not easy to see how the atmosphere would acquire this by its own internal adjustments. This points to an external agency, though perhaps not external to the oceans and atmosphere if these are considered as a joint system (see § 10.1.1).

Temperature is the best-documented of the weather elements and the easiest to infer from other evidence. Climatic change is not, however, a matter of temperature fluctuation alone; a change in the nature of the general circulation is implied and therefore, also, in the distribution of rainfall. For those many parts of the world which are marginally placed with respect to rainfall, a trend in this element is of even more vital concern than is a change in temperature anywhere.

The causes of climatic change that have been proposed are almost too numerous to detail. They include changes of solar radiation, earth-orbit fluctuations, mountain building, continental drift, volcanic eruptions, changes of the carbon dioxide and ozone content of the atmosphere, interactions between atmosphere and oceans, and atmospheric pollution. A more-or-less plausible case can be made for each of these, and they may indeed all be effective to some extent. However, no real concensus of opinion among research workers in this subject seems yet to have emerged as to which of these or other factors may be dominant.

A summary of this kind does much less than justice to the volume of expert knowledge that exists in relation to climatic change, in all its aspects, through the ages and in modern times. Our concern in this chapter is, however, with the forecasting problem: what is the probable trend of climate, in the British Isles or elsewhere, during the next few decades, say?

Climatological statistics have not revealed any periodicity on which a prediction may be based but, together with other indications, they suggest that there has been a recession in temperature, widespread but not world-wide, since the pre-1940 warming. Since the cause of this recent trend, any more than that of previous trends, is not known, there is no way of determining whether its reversal is imminent or whether it will continue. Statistical investigation has shown that, on average, the best forecast of rainfall and temperature for a few decades ahead is one based on the averages for about the past 15 years, this being a period long enough to reduce the importance of individual ' freak ' years but

223

short enough to exclude older climatic records which are no longer relevant. For sub-polar regions the more pessimistic—some would say the more provident—forecast would be for a continuation of the cooling trend. It is clear, however, that improvement in forecasting climatic trends must await better basic understanding of the interplay of atmospheric processes.

11.5.4. *Weather and climate modification*
Considerable thought and effort have been devoted, since about 1946, to the possibilities of modifying weather and climate by methods, and on a scale, beyond those which were already established in horticulture and agriculture. Consideration of these possibilities did, in fact, provide part of the stimulus for the Global Atmospheric Research Programme referred to in Chapter 7.

One of the fundamental facts of meteorology is the vastness of the energies involved in weather processes compared to any that man can bring to bear. Of many possible illustrations, it suffices perhaps to say that the total energy released in a nuclear explosion of moderate power is about equivalent to that of an average thunderstorm and is only about 10^{-5} that of an average depression. It is thus obvious that if man is to modify weather to a significant extent it cannot be by direct injection of energy but by more subtle means†. Small-scale illustrations are the practices employed in the protection of valuable crops against damage by radiation frost: namely ' smudging ' (in which the production of a dense smoke reduces radiative cooling of the surface), and also flooding the ground with water (the latent heat continuously released on freezing preventing ground temperature from falling much below 0°C).

Since 1946, most of the efforts at modification have been directed towards altering the natural processes of condensation and precipitation, generally with the aim of producing artificial rain. The possibility of so doing was first demonstrated in the U.S.A. by a simple experiment. Crystals of solid carbon dioxide (which is called dry ice and has a temperature of about $-80°C$) were dropped into a vessel containing a cloud of supercooled droplets and were seen to produce tiny ice crystals; these grew rapidly and fell out, while the original cloud also disappeared. This was, in fact, a direct demonstration of the Bergeron process described in Chapter 4. It was followed by the further demonstration that the dropping of dry ice from an aircraft into a natural supercooled cloud had the same effect.

Variations of the method of cloud seeding have been tried, in which silver iodide crystals (dropped from aircraft, or generated at the ground as fine smoke, or fired by rocket) have replaced dry ice as the seeding

† An exception was the emergency clearing of fog, along airfield runways in England in World War II, by direct heating of the air. These measures showed that very large amounts of fuel were required to effect even a partial and temporary clearance over a small area.

agent, silver iodide being used because its crystal structure is very similar to that of ice. Also, ' warm ' clouds (i.e. clouds at a temperature above 0°C) have been seeded by particles of common salt, or by water drops, in an attempt to stimulate the coalescence mechanism of precipitation within them. Besides rain-making, the experiments have been concerned with the clearance of (suitable) layer cloud and with the suppression of large hailstones which, on occasion, cause great damage. In the case of hail, the idea is to provide so many condensation nuclei that the cloud water is shared by much more numerous but smaller hailstones, most of which may melt before reaching the ground, than would otherwise be the case.

The results of seeding supercooled layer cloud tend to be clear-cut because, if successful, the clearance in the seeded area appears in a relatively short time. With suitable cloud droplets and seeding technique the method is effective and is used at times to clear ' cold ' (i.e. supercooled) fogs from airfields—these being a minority of the fogs that occur in Britain, even in winter. Seeding at frequent intervals is, of course, normally required to maintain such a clearance.

With rain-making, hail-suppression, and other similar attempts at modification, the results have been found very much more difficult to assess. The reason is that, in trying to promote rain, for example, it is only with clouds which are capable, or nearly capable, of producing rain by *natural* processes that seeding has any prospect of success. Since the distribution of natural rainfall in a given area is usually irregular, it is difficult to decide whether or not a given pattern of rainfall demonstrates an effect of seeding. It should be noted that, from the physical viewpoint, a large influence is unlikely. Rainfall amount is determined by vertical motion and by the amount of available water and its distribution among droplets. Seeding does not change the amount of water available but may, of course, affect its distribution; vertical motion may also be modified by the latent heat which is released after seeding.

The considered view (published in 1966) of an expert body of American meteorologists, who took into account the results of seeding in various countries, was that general rain of the orographic type, and also rain from cumulus clouds, are augmented by about 10%, on average. The results of other types of seeding were considered inconclusive.

It is not difficult to list a variety of actions which, if put into effect, would certainly cause changes of climate on a world-wide scale. Suggestions have included alteration of the albedo of the earth's surface, control of evaporation rate from the oceans, modification of the geography at selected sensitive points, deflection of ocean currents, interference with large-scale dynamics by appropriate triggering of disturbances— also various others of greater or lesser feasibility. Mostly, the requirements are far from being practical propositions; in no case is basic knowledge of cause and effect in large-scale meteorological processes yet sufficient for any major measure to be embarked upon.

It need hardly be added that, if the meteorological problems and those of physical means are solved, the ensuing political problems will prove to be at least as difficult. Whether issues of this kind will arise remains to be seen.

ANSWERS TO PROBLEMS

2.1. $\bar{M} = 28 \cdot 9$ kg kmol^{-1}; $\bar{M} = 22 \cdot 8$ kg kmol^{-1}.

2.2. For N_2, $c/v_e = 0 \cdot 33$; for 0, $c/v_e = 0 \cdot 43$; refer to Table 2.1.

2.3. $r = 20 \cdot 4$ g kg^{-1}; $e = 31 \cdot 8$ mb; $e = 40 \cdot 3$ mb.

2.4. $f = 37\%$; $f = 11\%$.

3.1. $\pm 3 \cdot 4\%$; identical.

3.2. (a) $1 \cdot 53 \times 10^{-2}$ W m^{-2}; (b) $0 \cdot 85 \times 10^{-2}$ W m^{-2}; (c) $0 \cdot 35$ kW m^{-2}; mean temperature $= 275$ K.

3.3. 22%.

3.4. $20°C$; $2°C$; air has warmed by $1 \cdot 5$ K by condensation of water vapour.

3.5. $3 \cdot 5$ cm s^{-2}; $5 \cdot 2$ cm s^{-2}.

4.2. 10%.

4.3. When $dr_W/dT \approx 0 \cdot 4$, i.e. at $7 \cdot 5°C$.

5.2. 2720 m (arctic); 2900 m (tropical); 740 m (frontal zone).

5.3. At $T = 20°C$ and $T_d = 16°C$, water content is $4 \cdot 0$ g m^{-3} at 700 mb and $5 \cdot 4$ g m^{-3} at 500 mb; $T = 20°C$, $T_d = 12°C$, $3 \cdot 8$ and $4 \cdot 3$ g m^{-3}; $T = 25°C$, $T_d = 16°C$, $3 \cdot 1$ and $5 \cdot 0$ g m^{-3}; $T = 18°C$, $T_d = 16°C$, $4 \cdot 5$ and $5 \cdot 6$ g m^{-3}.

5.4. r (g kg^{-1}): 4; (1) 4; (2) 4. U (%): 75; (1) 97%; (2) 57%.
e (mb): $5 \cdot 8$; (1) $5 \cdot 5$; (2) $6 \cdot 1$. θ (°C): 12; (1) 12; (2) 12.
T (°C): 1; (1) $-1 \cdot 5$; (2) 4. θ_W (°C): 6; (1) 6; (2) 6.

5.5. $\Delta T = 0$K (dry adiabatic), 4K (saturated adiabatic), $10 \cdot 5$K (isothermal).
$\partial T / \partial t = -w(\Gamma - \gamma)$, where w is vertical velocity and γ is environmental lapse rate.

5.6. $2 \cdot 2$ mm hr^{-1}.

6.1. π m in advance of target.

6.2. $\pi \sin \phi / 72$ m $= 3$ cm for $\phi = 45°$. Deviation is 60 times less for rifle bullet.

6.3. $\pi / 4$ km.

6.4. $1 \cdot 54$ cm; 52 m s^{-1}.

8.1. (V_G) cold $= 33 \cdot 3$ m s^{-1}; $V_F = 16 \cdot 7$ m s^{-1}.

8.2. (V_G) warm $= 25 \cdot 0$ m s^{-1} from direction 312°.

8.3. $\zeta(r) = +0 \cdot 2$ hr^{-1}; $\zeta(<r) = +0 \cdot 4$ hr^{-1} if $\partial V / \partial r = V/r$; $\zeta(>r) = 0$ when $Vr = $ constant as this implies $\partial V / \partial r = -V/r$.

8.4. $-0 \cdot 08°C$ per 3 hours.

8.5. $11 \cdot 5$ mm hr^{-1} (note, 1 kg m$^{-2} = 1$ mm, for rainfall).

9.1. $z_0 = 54$ mm.

9.2. C_D (10 m) $= 0 \cdot 0059$; $\tau = 0 \cdot 30$ N m^{-2}; $\tau = 0 \cdot 13$ N m^{-2}.

227

9.3. ($z_0 = 16$ mm; C_D (10 m) $= 0 \cdot 0039$); $\tau = 1 \cdot 87$ N m^{-2}.

9.4. Ri $= -0 \cdot 026$; $z_0 = 23$ mm; $u_* = 0 \cdot 52$ m s^{-1}; $\tau = 0 \cdot 32$ N m^{-2}; $H = 0 \cdot 22$ kW m^{-2}; $E = 2 \cdot 0 \times 10^{-4}$ kg m^{-2} s$^{-1} = 0 \cdot 72$ mm hr^{-1}; $LE = 0 \cdot 49$ kW m^{-2}; $B = 0 \cdot 45$; approximate surface values are $23 \cdot 2°C$, 70%, $20 \cdot 2$ mb and 475 mg m^{-3}.

9.5. K_M (10 m) $= 2 \cdot 1$ m^2 s^{-1}.

INDEX

The more important references are in italics the less direct in parentheses.

Absorption 7, 23, *25–32*, 100, 111, 113
Adiabatic processes 33–36
Adiabatic saturation temperature (see Temperature)
Advection 48, *94–96*, 144, 162, 213
  differential 96
  temperature 94–96
Aerodynamic roughness (see Roughness)
Agriculture 4, 222, 224
Air 5–8
Aircraft 104, 107, 204, 224
Air density 6, *9*, 109, 110
Airfield 104, 205, 224
Air mass (96), *123–129*, (134), 136, 156–158, 200
Air mass analysis 104, 157, 158, 204
Air mass characteristics 74, *126–128*
Air mass classification 75, *124–128*
Air mass modification 127, 210
Air pressure 9
Air temperature (see Temperature)
Albedo *26*, 39, 40, 50, *117*, 185, 225
  planetary 27
  visual 27
Aleutian low 187–189
Alps 52, 142
Altimeter equation *10*, 65, 86, 100
Altocumulus 51, 53, *55*, 56. Plates 3, 4, 5 and 11
Altocumulus castellanus 128
Altostratus 50, *55*, 56, 133. Plate 9
Anabatic wind (see Wind)
Ana cold front 134
Analogue methods (of forecasting) *216*, 217
Anemometer 99, *103*, 164, 173
  hand 114, 116, 118
  sensitive cup 114, 116
Angular momentum 149, 197, 198, 200, 201
Anomalous audibility (see Audibility)
Antarctica 126. Plate 5
Anticyclone 3, 71, 88, 90, *122–124*,

127, *143–145*, 147, 148, 151, 153, 194, 199, 213, 221
  cold 144, 145
  sub-tropical 137, 142, 145, 186, *188–190*, *200–202*
  warm 144, 145, 209
Anticyclonic curvature (see Curvature)
Anticyclonic gloom 144
Anticyclonic spin (see Spin)
Anti-trades 201
Anvil cloud 49, 57. Plate 2
Aphelion 24, 40
Arctic 126
Arctic front 134, 135, 200
Arctic sea smoke 48
Argon 5
Artificial rain (see Rain)
Asia 190, 202
Atlantic 124, 136, 142, 143, 156, 158, 187
Atmosphere, standard 64
Atmospheric pollution (see Pollution)
Atmospheric pressure (see Pressure)
Atmospheric window 30, 31
Audibility 108
  anomalous 108
Automatic Picture Transmission (A.P.T.) 207, 208. Plate 13
Automatic weather station (99), 112, 113
Aviation 81, (107), 203, 204
Azores 122

Backing (of wind) *90*, 91, 164
Balloon 59, 104, 106, 107, 110, 112, 118, 204
  constant-level 98, 113
Banner cloud 54. Plate 6
Baroclinic model 220
Barograph *100*, 114, (210), (212)
Barometer 9, 99, 100, 105
Barometer equation 9
Barotropic model 217–220
Bath plug vortex 149

229

Beaufort force 105, 142, 159, *161*, 162
Beaufort scale 105, 142, 161
Bergeron process *44*, 224
Billow cloud 57
Black ice 47
Blocking 155
Blocking high 156
Boundary layer, laminar 165, 175, 182
  planetary 164, 173, 174
  surface 164
  turbulent 2, 164, *165–182*
de Bort, Teisserenc 106
Bowen's ratio 177, 181
Brewer, A. W., 111
British Broadcasting Corporation (B.B.C.) 157, 159, 160
British Isles 127, *128*, 134, 143, 156, 203, 208, 210, 212, 213, 216, 222, 223
British Meteorological Office 109, *157*, 206, 216, 221
Brush discharge 46
Buoyancy 37, 38, 49, 179, 180
Buys Ballot's law 84

Campbell-Stokes sunshine recorder 104
Canada 145, 158
Capetown Plate 7
Carbon dioxide 2, 5, *29*, *30*, 113, 181, 223
  solid 224
Castellanus 50, 58, 128
Characteristic curve 75
Circle of inertia 83
Circulation indices 155
Circumpolar vortex 190, 201
Cirrocumulus 55. Plate 12
Cirrostratus 55, 56, 133. Plates 1 and 3
Cirrus 43, 55, 57, 133
  false 55
Climate 184, 201, 222
Climatic change (trends) 222–224
Climatic zones 201, 202
Climatological station 98
Climatology 98, 99
Cloud 21, *41–46*, *49–58*, 132, 137
  high 55, 57, 119
  lenticular 53. Plates 4 and 5
  low 55, 104, 119
  medium 55, 57, 119
  mixed 44, 55

noctilucent 107
orographic 52, 53. Plate 7
supercooled (45), 224, 225
warm 225
Cloud amount 30, 115, 119, (122), 158, 185, 208, 212
Cloud base 56, 73, 104
Cloud chamber 41
Cloud droplets 41–45, 58, 224, 225
Cloud genera *55*, 104. Plates 1–12
Cloud photography, infra red 113
Cloud searchlight 104
Cloud seeding 224, 225
Coalescence *44*, 225
Col *123*, 157
Cold fog (see Fog, supercooled)
Cold front clearance 137. Plate 11
Cold front wave 137, (138), (162)
Cold sector 136
Cold trough 155
Computers (see Electronic computers)
Concrete minimum temperature (see Temperature)
Condensation *41–54*, *62*, 74, 75, 142, 143, 224
Condensation level, convective (CCL) 67, *73*
  lifting (LCL) 52, *67–77*
  mixing 51, 67
Condensation nuclei 18, *41–43*, 225
Conductance 172, 175
Conduction 6, 32, 33, 38, 39, 47, 48, 177
Conductivity 39, 174
Confluence 146
Conservation laws 196–198
Conservation of angular momentum 149, 197
Conservative properties (of air masses) 74
Constant level analysis 86
Constant-level balloons (see Balloon)
Constant pressure analysis (see Isobaric analysis)
Continental drift 2, 223
Convection 23, 29, *31–38*, 52, 72, 111, 175, 184
  forced 32, 46, *180*
  free (natural) 32, 176, 180
Convection, level of free (LFC) 72, 73
Convective condensation level (CCL) (see Condensation level)
Convective equilibrium 35, 38

Convergence   71, 125, 140, 142, 143, *145–154*, 163, 201, 209
frictional   125, 148
Coriolis force (C.f.)   14, 78, *80–90*, 129–132, 149, 199
Coriolis parameter   *82–94*, 149–154, 218, 219
Critical angle of incidence   108
Cumulonimbus   45, 46, 49, 50, 55–57, 128, 132, 133, 140, 142, 143, 199. Plate 2
Cumulus   43, 49, 50, 52, 55, 56, 73, 128, 132, 133, 225. Plates 1, 3, 9 and 12
trade wind   201
Cumulus street   50. Plate 3
Curvature (of flow)   88, 89, *127*, (149), 150, 151, 154
Curvature (of isobars)   88, 127, 159
Curvature, anticyclonic   *88*, 89, *127*
cyclonic   *88*, 89, *127*, 150
Cut-off low   156
Cyclone   145, *151*, 190, 200
Cyclonic circulation   90, (127), 142, 143
Cyclonic curvature (see Curvature)
Cyclonic spin (see Spin)
Cyclonic vorticity (see Vorticity)
Cyclostrophic term   89

Daily Weather Report (D.W.R.)   153, *157*
Damping (of turbulence)   180
Day-length   197
Depression   3, 71, 122, 123, *135–143*, 147, 148, 151, 153, 156, 158, 194, 199, 203, 208, 209, 212, 221, 224
frontal   *134–139*, 142, 158
lee   141
non-frontal   139–143
old   157
parent   137
polar   140, 141
primary   137, 139
secondary   137, 139
warm-sector   136
Depression family   137, 138
Dew   20, *46*, 47
Dew-point temperature (see Temperature)
Differential advection (see Advection)
Diffluence   146
Diffusion, molecular   5, 46, *166*, (167), 173, 175
turbulent   2, *166*, 172, 173

Diffusive equilibrium   10
Diffusivity, molecular   22, 182
turbulent   182
Dines meteorograph   106
Dishpan experiment   195
Dissociation   7, 8
Divergence   71, 123, 124, 142, *145–154*, 209
Dobson, G. M. B.   109, 111, 193
Doldrums   88, 201
Drag, aerodynamic   103, 169, 171, 175
Drag coefficient   171, 172, 175, 181
Drizzle   *44*, 52, 120, 127, 128, 133
freezing   47
Dry adiabatic lapse rate (DALR) (see Lapse rate)
Dry adiabatics   59, *60–77*
Dry ice   224
Dust   26, 41, 108, 185
Dynamical modification (of air masses)   127

Earth temperature (see Temperature)
Economics   4, 214
Eddies   32, 54, 102, 103, 105, 118, *167*, 201
Eddy velocity   167
Eddy viscosity (see Viscosity)
Electricity   41, 45, 105
Electronic computers   183, 206, 216, 220
Emission   23, *28*, 29, (113)
Energy budget   105, 197
Entropy   60
Environmental lapse rate (see Lapse rate, environmental)
Environment curve(s)   37, *64–77*
Equator   15, 79–81, 83, *87*, 88, 106, 142, 145, 186, 193–201
Equatorial zone (region)   15, 184, 196
Europe   159, 204, 206, 216
Evaporation   15, 19, 20, 39–43, 46, 48, 72, 74, 127, 133, 142, 158, 173, 192, 197, 198, 222
Evaporation rate   115, 116, (175), 177, 181, 185, 225
Evaporation tank   103, (115)
Exosphere   8, 13, 21
Explosion   108
Explosive   8, 13, 21
Explosive warning   194
Exposure (of instruments)   99, 102, 103, 105
Eye (tropical cyclone)   142, 143, 206

Fallout, radioactive 195
Farman, J. C. Plates 5 and 6
Fiducial point 100
Field experiments 99, *114-118*, 173
Floods 213
Flux, turbulent 166, 171, 172, *174-177*, 181
Fog 42, 46, 51, 120, 122, 127, 174, 208, 210, 221, 224
  advection 48, 144
  coastal 128, 144, 213
  frontal 49
  ground 47
  hill 103, 128
  radiation 47, 52, 213
  sea 48, 128
  steam 48, 52
  supercooled 46, 47, 225
Föhn 52, 53, 67, *68*, 128
Forecast 102, 119, *203-222*
Forecast chart 157, 204, 210
Forecaster 4, 155, *205-221*
Forecasting 4, 98, 99, 107, 137, 153, *203-220*
  local 203, 205, (221)
  long-range *213-217*, 221
  numerical 153, 206, 217-221
Forecasting equation 217, *218*, 220
Forecast period 159, 204, 210, 215-217
Franklin, Benjamin 105
Friction 14, 81, 85, *90*, 92, 97, *122*, 124, 125, 136, 137, 148, 164, 197, 220
Friction layer 50, 51, 78, 85, *90-92*, *164*
Friction velocity 167-182
Front 54, 94, 96, 97, 120, 121, *124-142*, 152, 158, 162, 204, 208, 210
  cold 120, *125*, 126, 132, *134-137*, 139, *141*, 145, 157-159, 162. Plate 11
  occluded 120, 125, *135*, (136), (139)
  quasi-stationary 120, 125, *135*
  warm 55, 120, 125, 126, *132-137*, 139, 158, 159
Frontal analysis 157-159, 204
Frontal depression (see Depression)
Frontal jet (see Jet stream)
Frontal location (157), *158*
Frontal passage 137, 156, 212. Plate 11
Frontal slope 76, *124-126*, *130-134*, 162
Frontal speed 125, 126, 133, 162

Frontal structure 132-134
Frontal surface 49, 76, 96, *125*, 126, *129-134*
Frontal trough (see Trough)
Frontal wave 135, 136, 141, 142, (162)
Frontal wind shear (see Wind shear)
Frontal zone 76, *124*, 125, 130, 133, 134
Frost 45-47
Frost (below freezing temperatures) 128, 144, 210, 224
Frost, glazed 46, 47
  ground 102
  hoar 46, 47
Frost point 192
Funnel cloud 143

Gales 213
Gamma rays 23
Gates, D. M. 30
General circulation 3, 123, *183-202*
Geography 199, 225
Geology 2, (222)
Geomagnetism 2
Geostrophic balance *84*, 87, 88, 122, (131)
Geostrophic flow (or motion) *83*, 126, *130-132*, 146, 164, 218
Geostrophic force 82
Geostrophic speed 151
Geostrophic wind (see Wind)
Geostrophic wind equation 84-86
Geostrophic wind scale 85, 86, 94, 161
Germany 206, 212, 215
Glaciers 223
Glazed frost (see Frost)
Global Atmospheric Research Programme (GARP) 112, 113, 224
Global Horizontal Sounding Technique (GHOST) 113
Gold-beater's skin 110
Gradient wind (see Wind)
Grass minimum temperature (see Temperature)
Gravitation 209
Gravity (g) 14, 15, 80, 86, 99
Greenland 124, 163, 212
Grid points 219
Ground frost (see Frost)
Gulf of Lyons 142
Gust 141, 221
Gustiness 91, 170, 221

Haar 128
Hadley 183, 198, 199
Hadley cell 198, 199
Hail 45, 56, 225
Hail showers (56), 120, 128, 140
Hail-suppression 225
Hailstones 45, 225
Hair hygrograph 102
Halo 55, 133
Hand anemometer (see Anemometer)
Heat budget 105
Heat capacity 39, 185
Heat engine 183
Heat high 140
Heat low 139, 140, 190
Hedgerow 116, (171)
Helium 7, 8
Hill fog (see Fog)
Historical surveys 18, 105–109, *203–208*
Hoar frost (see Frost)
Hodograph 94–96
Horse latitudes 201
Horticulture 224
Howard, Luke 55
Humidity, relative *19*, 21, 41, *65*, 74, 101, 102, 110, 114, 115, 118, 127, 143, 181
Humidity profile 75, 115, 182
Humphreys, W. J. 108
Hurricane 14, 142, 143
Hydrogen 7, 8
Hydrology 2
Hydrostatic equation 10, (35), 68
Hygrograph 102
Hygrometer 20
Hygroscopic nuclei 42, 47

Ice 47, (101), 108, 169, 208, 216, 225
Ice Ages 185, 222
Ice crystals 17, *43*, 44, 49, 55, 57, *224*
Iceland 140, 153, 212
Icelandic low 187–189
Ice nuclei 43
India 190, 202
Inertial flow 83, 89
Infra-red cloud photography (see Cloud)
Infra-red radiation (see Radiation)
Instability *36–38, 71–74*
  conditional 38
  convective (see potential)
  latent 72, 73
  potential 71–73, 77, 134

Inter-tropical convergence zone (ITCZ) 142, *186–189*, 200, 202, 204
Inversion (temperature) *11*, 32, 46, 47, 51, 53, 56, 71, 72, 108, *144*
Inversion conditions 178, 180
Ion 7, 45, 46, (112)
Ionization 7, 25, (112)
Ionosphere 2, 7, *11*, 25, 46
Isobar 59, *61–63, 83–85, 162, 203*
Isobaric analysis 86, 110
Isobaric layer 10, (64), (68–71), 139, 206
Isobaric surface (86), 87, *92–94*, (110), 152, 153, 198, 217
Isobaric thickness (10), *65*, 76, *93, 94*, 137, 139, 155, 156, 204, 206, 216
Isobaric trough (see Trough)
Isobar spacing 84, 85, 122, 170
Isotherm 59, 60, *62–76*, 92, 93, 123

Jet stream 54, 55, 78, *96*, 97, 134, 135, 151
Jet stream, frontal 97, 134, 135
  polar front 151, 204
  polar night 97
  sub-tropical 97, 200, 201
Jet stream cirrus 57, 204
Jet stream levels 153, 204

von Kármán's constant 168
Katabatic wind (see Wind)
Kata cold front 134
Kinematic viscosity (see Viscosity)
Kinetic energy 136, 137, 179, 180
Kirchoff's law 29
Kites 105, 106

Lamb, H. H. 128
Laminar boundary layer (see Boundary layer)
Laminar sub-layer 166, 167, 169
Laminar flow 165, 180
Land breeze 40
Land–sea distribution 185, *202*
Lapse conditions 11, 178
Lapse rate 11, 23, 29, 33, *35–38*, 51, 52, 59–77
  dry adiabatic (DALR) *35*, 51, 52, 59–77
  environmental 37, 38, 70, 71
  mixing ratio 51
  saturated adiabatic (SALR) *36*, 51, 52, *62–77*

233

Latent heat   15, 19, 23, 31, *36*, 45, 47, *62*, 101, *142*, 177, 197, 220, 224, 225
Leakage current   46
Lee depression (see Depression)
Lee trough   141, 151
Lee waves   53, 56. Plate 5
Lenticular cloud (see Cloud)
Level of free convection (LFC) (see Convection)
Level of non-divergence   *154*, 217, 218, 220
Lightning   45, 99
Lightning conductor   106
Lifting condensation level (LCL) (see Condensation level)
Lifting path curve   72, 73
Lindemann, F. A.   109
Local forecasting (see Forecasting)
Long-range forecasting (see Forecasting)
Long waves   *154, 155*, 196
Lorenz, E. N.   221
Ludlam, F. H.   54
Lunar day   209

Mars   21
Maximum temperature (see Temperature)
Maximum tnermometer (see Thermometer)
Mean free path (molecular)   166, 167
Mediterranean climate   202
Meridional flow (or motion)   *155*, 197, 200
Mesopause   11, *12*, 190, 192
Mesosphere   *11*, 12, 29, 30, 32, 195
Meteor   107, *109*
Meteorite   109
METMAP   *157, 159–162*
Microbiology   3
Microclimate   178
Minimum temperature (see Temperature)
Minimum thermometer (see Thermometer)
MINTRA (lines)   64
Mintz, Y.   192
Mist   120, 128, 161, 174
Mixing condensation level (see Condensation level)
Mixing length   166, *167*, 168, 178, 179
Mixing ratio (humidity)   *16*, 21, 36, 51, 52, *61–69*, 74, 77, 192
Mixing ratio lines   61–77

Molecular weight   *6*, 16, 21
Momentum   32, *171*, 175, 182, 185
Momentum flux   *166*, 167, 172
Monsoon low   202
Monsoon winds   40
Moon   27, 40, 209

Nephoscope observations   204
Newton's laws   78, 79, 81
Nitrogen   5, 7, 8, 21
Nimbostratus   50, 55, 56, 133. Plate 10
Nimbus   55
Noctilucent cloud (see Cloud)
Non-frontal depressions (see Depression)
Normand's theorem   *66*, 74
North America   190, 202, 203
Nuclear explosion   194, 195, 224
Numerical weather forecasting (see Forecasting)

Occlusion   135, 136, 157
  cold   135, 136, 139
  warm   135, 136, 139
Occlusion (process of)   136
Occlusion, secondary   139
Ocean currents   32, 225
Oceanography   2
Oceans   185, (202), 209, 223, 225
Ocean Weather Ships (O.W.S.)   98, 119, 122, 204
Ohm's law   174
Orographic cloud (see Cloud)
Orographic low   *141*, 142, 154
Orographic rain (see Rain)
Orographic throug   141, 154
Oxygen   5, 7, 8, 21, 27
Ozone   5, 25, *27*, 30, 98, *111*, 112, *193*, 194, 195, 223
Ozone sonde   111, 112
Ozone spectrophotometer   111

Pacific   142, 187
Parent depression (see Depression)
Perihelion   24, 40
Periodicities   214
Photochemical equilibrium   193
Photosynthesis   2
Planetary boundary layer (see Boundary layer)
Plant physiology   3
Plotting   159
Plotting code   119, 120, 161
Polar front   *134–138*, 200, 201

Polar front jet stream (see Jet Stream)
Polar low   140, 141
Polar night jet (see Jet stream)
Politics   226
Pollution, atmospheric   2, (47), *144*, 159, 173, *174*, 223
Position circle   122
Potential energy   136
Potential instability (see Instability)
Potential temperature (see Temperature)
Potential vorticity (see Vorticity)
Planck   24
Precipitable water   68, 222
Precipitation   15, *41–45*, (50), *102*, *103*, 128, 134, 137, 195, 197, 198, 202, 222, 224, 225
Precipitation–evaporation cycle   222
Precipitation rate   *68*, 69, 77
Press, the   119, 157, 203
Pressure, atmospheric   5, *9–13*, 99, *100*, 188, 189, 199, 203, 211, 212, 214, 216, 221
Pressure forces   171, 175
Pressure gradient   *14*, 78, 79, 85, 92, 131, 143, 209
Pressure gradient force (p.g.f.)   *14*, 15, 78–80, *83–90*, 92, 129, 130
Pressure tendency   114, 119, 158, 162, *208*, 209, 210
Profile experiments   114, 115, 117
Psychrometer   20, 22, 101
   whirling   114, 118
Psychrometer constant   20, 22

Radar   99, 204, *205*
Radar winds   109–111, 204
Radiation   6, 20, *23–32*, 46, 74, 104, 111, 144, 145, 159, 185, 190, 192, 196
   black body   23, 28, 29
   diffuse (solar)   26, 28, 105, 117
   direct (solar)   26, 28, 105, 117
   infra-red   23–31, 113
   long-wave   *29*, 31, 112, 184
   net   (31), 46, (105), *177*, 194, 200
   short-wave   *29*, 104, 105, 117
   solar   *23–32*, 40, 100, 105, 110, 117, 184, 187, 199, 223
   terrestrial   24, (28)
   ultra-violet   7, 8, 23–27, 111
   visible   23, 24
Radiation balance   31, 32
Radiation inversion   46, 72, 144
Radiation night   29, (72)

Radiation shield   20, 110, 115
Radiative cooling   47, 145, 159, (180), 190, 202, (210), 224
Radiative equilibrium   32
Radio   99
Radioactive tracers (see Tracer)
Radiometer   26, 105
   net   105
Radio propagation   7
Radiosonde   86, 107, *109*, 110, 112, 204, (219)
Radio waves   7, 23
Rain   5, *45*, 46, 72, 74, 120, 133, 134, 142, 162, 173, 174, 205, 213
   artificial   224, 225
   freezing   46
   orographic   52, 69, 225
Raindrop   42, 44, 45, 49, 102, 103, 205
Rainfall   52, 69, *102*, 103, *114–116*, 163, 185, 201, 202, 209, 211, 212, 214, 216, 217, 222, 223, 225
Rain gauge   102
Rain-making   225
Rain shadow   52
Rain showers   120, 140
Relative humidity (see Humidity)
Representativeness (of observations)   158, 159
Research   4, 183, 221, 222
Resistance   *174–182*
Resistivity   182
Return eddy (see Eddies)
Richardson, L. F.   3, *206*
Richardson's number   180, 181
Ridge   *122*, 141, 151, 153, 156, 158, 199, 210, 212
Rime   46, 47
Rockets   107, 108, 112, 224
Rocket sonde   112
Roaring Forties   201
Rocky Mountains   190
Rossby waves   155
Rotation, anticyclonic   88, 90
   cyclonic   88, 90
Rotor cloud   54
Roughness, aerodynamic   *168*, 169, (171), (172), 175, (185)
Roughness length   117, *168–170*, 172, 177, 178, 181
Royal Meteorological Society   *157*, 216

Sahara   126
Salt   41, 225

Sand 117, 169
Satellite 23, 27, 98, 112, 113
Satellite photography (113), 142, 207, 208. Plate 13
Saturated adiabatic lapse rate (SALR) (see Lapse rate)
Saturated adiabatics 59, *62*, *63*, 66
Saturation vapour pressure (s.v.p.) (see Vapour pressure)
Sawyer, J. S. 132, 221
Scale height 13, 14
Scattering 23, *25*, *26*, 28
Scherhag, R. 191
Scorer, R. S. 54
Scottish Highlands 128. (Plates 8 and 11)
Sea breeze 40, *140*
Secondary depression (see Depression)
Seismology 2
Sellers, W. D. 31
Shearing stress *165–167*, 171, 181, 182
Sheppard-type cup anemometers (see Anemometer, sensitive cup)
Shipping forecasts 157, 159, 160
Showeriness 127, 209, *221*
Showers 56, 72, 120, *128*, 140, 159, 201, 212, *221*
Showery precipitation *49*, (56), 133, (140), 201, 202, (205)
Siberia 145, 202
Silver iodide 224
Sleet 45
Sleet showers 128
Smagorinsky, J. 221
Smog 47, 144, 174
Smoke 47, 173, 224
Smudging 224
Snow 26, 29, 39, *103*, 120, 122, 128, *140*, 169
Snow cover 123, 159, (207), 208
Snowflake 45, 103
Snow showers 120, 128, *140*
Snow-storm 153, 213
Soil 27, *39*, 117
Soil physics 3
Soil temperature (see Temperature)
Solar constant 24, 25
Solar day 209
Solarimeter 105, *114*, 117
Solar variability 25, 215
Sound propagation 107–109
Source regions 123, 124, *126*
Spin 149, 153, 220
  anticyclonic 149, 154
  cyclonic 140, 143, 149, 154

Spitzer, L. 8
Spray 41, 105
Squalls 140
Stability 33, *36–38*, 68, *70–74*, 96, 126, 127, 133, 178, 210
  neutral 38, 178
Standing waves 190
Statistics 214–221
Stefan's law 29, 30
Stratocumulus 51, 53, *55*, 56, 128, 133, 144. Plate 8
Stratopause 11, 12, 190
Stratosphere *11*, 12, 20, 27, 29, 30, 32, 53, 97, 106–108, 134, 147, 183, 190, 192–195, 214
Stratus 43, 47, 48, 51, *55*, 56, 128, 133, 144, 208, 210. Plates 7 and 11
Streamlines 139, 146
Streamlining 171, 181
Subsidence 36, *71*, 77, 133, (143), *144*, 146, 148, 194, (201), 209
Sub-tropical anticyclones (see Anticyclone)
Sunshine 100, 104, 187, 216
Sunshine recorder 104
Sunspots 25
Supercooled droplets 43, 45, 224, 225
Supercooling 17, *43*, 45, 49
Supersaturation *17*, 41, 43, 44, 49, 51
Surface air temperature (see Temperature)
Surface analysis 156
Surface boundary layer (see Boundary layer)
Surface temperature (see Temperature, surface air or Temperature, surface)
Surface wind (see Wind)
Sutcliffe, R. C. 222
Sutton, O. G. 173
Synoptic chart 59, 79, *119–121*, 185, 186, 203, 204, 207–210, 212, 216
Synoptic codes 120, *157*, 161
Synoptic meteorology 104, *119–163*, 208
Synoptic station 98, 104

Telegraph (wireless) 203
Television 119, 157
Temperature 5, 9, *11–13*, 18, *28–43*, 100–109, 114, 185, 216, 217, 222, 223
  adiabatic saturation 19, 20
  air 6, *100*, 101, 118, 122, 140
  concrete minimum 102

236

dew-point *18*, 46, 48, *64–67*, 74, 77, 114, 115, 118, 158, 212
dry-bulb (20), 100, 101, 114, 115
earth 102
grass minimum 101
maximum 101, 115, 144
minimum 101, 102, 115, 221
potential *35*, 40, 60–77, 153
radiation 24, 113
radiative equilibrium 40, 184 (see equation (3.1))
screen 100, 101, *122*
sea-surface 43, 105, 140, 142, 158, 216
soil 102, 115
surface 48, 175, 178, 181, 184, 224
surface air 48, 72, 73, *122*, (140), 159, 162, 214
thermodynamic wet-bulb (Tw) *19*, 20, 66, 72, 74, 77, 158
wet-bulb potential *66*, 72, 75, 77, 192
virtual 17, 65, 111
Temperature advection (see Advection)
Temperature of a wet bulb (Tw′) *20*, 100, 101, 114, 115
Temperature gradient, horizontal 78, *92–97*, 123, 124, 201
vertical 164, 165, *178–181*
Temperature profile 75, 113, 115, *178, 179*, 182
Tephigram *59–77*, 208
Terminal velocity (see Velocity)
Thermal 32, 49, 52, 56
Thermal low 136
Thermal steering 137
Thermal wind (see Wind)
Thermal wind equation 93, (94)
Thermocouple 105, 115
Thermodynamics (first law) 34
Thermodynamic wet-bulb temperature (see Temperature)
Thermograph 102
Thermopile 104, 105
Thermometer 20, *100–102*
dry-bulb 101
mercury (-in-glass) 100–102, 115, 116
maximum 101
minimum 101, 115
wet-bulb 20, 22, 100, 101
Thermometer mounts 114, 115
Thermometer screen 22, *100*, 101, 102, 114

Thermosphere 11, 13
Thickness (see Isobaric thickness)
Thickness chart *93*, 155, 156
Three-cell model 199, 200
Thunder *45*, 56, 128, 134, 209
Thunderstorm 3, 41, *45*, 52, 120, 122, 140, 142, 213, *224*
Thundery outbreaks 72, (205)
Thunder cloud 105
Tidal motion 209, 210
Topography 154, 159, 202, 212, 220
Tornado 3, *143*
Tracers 74, 118, *190*, 192–194, 200
radioactive 194, 195
Trade winds 187, 199
Trade wind belts 201
Transfer coefficient (momentum) 171, 175
Transpiration 181, 197
Tree rings 222
Tropical cyclone *88*, (205), 212
Tropical meteorology 204, (210)
Tropopause 11, 12, 21, 29, 32, 49, 53, 92, 93, 96, 97, 106, 109, 142, 145, 190, 192–200
Tropopause break 133, 134, 193
Troposphere *11*, 12, 20, 21, 23, *29–32*, 78, *92–94*, 97, 106, 107, *143–147*, 151, 154, 183, *192–200*, 206, 209, 214
Trough *122*, 134, 141, 151, 153–157, 159, 212
frontal 125, 162, 210
Turbulence 2, *50*, 51, 144, *164–168*, 170, *173, 174*, 178–182
Turbulent boundary layer (see Boundary layer)
Turbulent mixing 52, 56, 57, (91), 125, 144, (173), (179), 194
Turf wall 103
Turn-table 81
Twister 143
Typhoon 142, 205

U.S.A. 126, 143, 153, 183, 215, 221, 224

Vapour pressure *16–22*, 48, 66, 75
saturation (s.v.p.) *17–22*, 42, 66, 108
surface 176, 178, 181
Vapour pressure profile 115, (182)
Vegetation 26, 39, 103, 117, (168), 171, 176, 177, (181), 185
Veering (of wind) 96

237

Velocity, escape 8
  terminal *42–44*, 58
Vertical motion 12, 33, (35–38),
  (44), 46, *49–54*, (62), *67–77*,
  84, 134, *145–147*, 190, 209, 210,
  225
Virtual temperature (see Temperature)
Viscosity 3
  eddy 182
  kinematic 182
Visibility 104, 119, 128, 158, 159
Vortex 149
Vorticity 145, *149–154*, 162, 217
  absolute *149–154*, 217, 218
  anticyclonic 149, 154
  cyclonic 149, 150, 151, 154
  geostrophic 217–219
  geostrophic absolute 218, 219
  potential 153, 154
  relative 149–150

Walker, Sir Gilbert 214, 215
Warm clouds (see Cloud)
Warm front wave 137, 139
Warm ridge 155
Warm sector *135*, 157, 209
Warm sector depression (see Depres-
  sion)
Water balance 103, 105
Water budget 197
Waterspout 143
Water vapour 8, *15–22*, 23, *25–32*,
  36, *41–54*, 113, 142, 143, 176, 177,
  185, 197
Wave clouds 53. Plate 5
Wave depressions *137–142*, 162,
  (210)
Waves (ocean) 105
'Weather' *157*, 216
Weather, local 145, *156–162*
Weather map (surface) 59, 79, *119–
  121*, 204
Weather radar *205*, 213
Weather ships (see Ocean Weather
  Ships)
Welsh, John 106
Westerly-type weather 201, *209–212*

Wet-bulb temperature (see either
  Temperature, thermodynamic
  wet-bulb (Tw) or Temperature
  of a wet bulb (Tw'))
Wet-bulb thermometer (see Thermometer)
Wexler, H. 54
Whirling psychrometer (see Psychrometer)
Wien's displacement law 28
Wilson, Dr. Alexander 105
Wilson, G. W. Plate 7.
Wind 14, *78–97*, 99, *103*, *116*, 142,
  143, 151, 158, 159, *164–181*, 185,
  186, 188, 189, 197, 204
  anabatic 52
  geostrophic *83–97*, 122, 125, 126,
    136, 162, 170, 217
  gradient *88–90*
  katabatic 52
  maximum gust speed (of) 99, 141,
    (143)
  surface (10 m) 90, 91, *103*, (119),
    122, (170)
  thermal 10, 78, *91–97*, 133, 137,
    139, 154, 184, 190, 199, 217, 220
Wind profile 117, 164, *165–182*
  logarithmic 165, *168–170*, 178–
    180
Wind rose 116
Wind shear 54, 56, 91, 151, 162
  anticyclonic 151
  cyclonic 135, (149), *150*
  frontal 130–135
  horizontal *149–151*, 162
  vertical 56, 57, *164–170*
Wind shear vector, vertical *56, 57,
  91–96*
Wind vane 103. Plate 10
World War I 107, 204
World War II 205, 224
World Weather Watch 112–114

X-rays 23

Zero-plane displacement 117
Zonal flow 155, 190
Zonal index 155
Zone of silence 108

THE WYKEHAM SCIENCE SERIES

1	*Elementary Science of Metals*	J. W. MARTIN and R. A. HULL
2	†*Neutron Physics*	G. E. BACON and G. R. NOAKES
3	†*Essentials of Meteorology* D. H. MCINTOSH, A. S. THOM and V. T. SAUNDERS	
4	*Nuclear Fusion*	H. R. HULME and A. McB. COLLIEU
5	*Water Waves*	N. F. BARBER and G. GHEY
6	*Gravity and the Earth*	A. H. COOK and V. T. SAUNDERS
7	*Relativity and High Energy Physics* W. G. V. ROSSER and R. K. MCCULLOCH	
8	*The Method of Science*	R. HARRÉ and D. G. F. EASTWOOD
9	†*Introduction to Polymer Science* L. R. G. TRELOAR and W. F. ARCHENHOLD	
10	† *The Stars: their structure and evolution*	R. J. TAYLER and A. S. EVEREST
11	*Superconductivity*	A. W. B. TAYLOR and G. R. NOAKES
12	*Neutrinos*	G. M. LEWIS and G. A. WHEATLEY
13	*Crystals and X-rays*	H. S. LIPSON and R. M. LEE
14	†*Biological Effects of Radiation*	J. E. COGGLE and G. R. NOAKES
15	*Units and Standards for Electromagnetism* P. VIGOUREUX and R. A. R. TRICKER	
16	*The Inert Gases: Model Systems for Science*	B. L. SMITH and J. P. WEBB
17	*Thin Films* K. D. LEAVER, B. N. CHAPMAN and H. T. RICHARDS	
18	*Elementary Experiments with Lasers*	G. WRIGHT and G. A. FOXCROFT
19	†*Production, Pollution, Protection*	W. B. YAPP and M. I. SMITH
20	*Solid State Electronic Devices* D. V. MORGAN, M. J. HOWES and J. SUTCLIFFE	
21	*Strong Materials*	J. W. MARTIN and R. A. HULL
22	†*Elementary Quantum Mechanics*	SIR NEVILL MOTT and M. BERRY
23	*The Origin of the Chemical Elements*	R. J. TAYLER and A. S. EVEREST
24	*The Physical Properties of Glass*	D. G. HOLLOWAY and D. A. TAWNEY
25	*Amphibians*	J. F. D. FRAZER and O. H. FRAZER
26	*The Senses of Animals*	E. T. BURTT and A. PRINGLE
27	† *Temperature Regulation*	S. A. RICHARDS and P. S. FIELDEN
28	†*Chemical Engineering in Practice*	G. NONHEBEL and M. BERRY
29	†*An Introduction to Electrochemical Science*	J. O'M. BOCKRIS, N. BONCIOCAT,
		F. GUTMANN and M. BERRY
30	*Vertebrate Hard Tissues*	L. B. HALSTEAD and R. HILL
31	† *The Astronomical Telescope*	B. V. BARLOW and A. S. EVEREST
32	*Computers in Biology*	J. A. NELDER and R. D. KIME
33	*Electron Microscopy and Analysis* P. J. GOODHEW and L. E. CARTWRIGHT	
34	*Introduction to Modern Microscopy* H. N. SOUTHWORTH and R. A. HULL	
35	*Real Solids and Radiation* A. E. HUGHES, D. POOLEY and B. WOOLNOUGH	
36	*The Aerospace Environment*	T. BEER and M. D. KUCHERAWY
37	*The Liquid Phase*	D. H. TREVENA and R. J. COOKE
38	†*From Single Cells to Plants* E. THOMAS, M. R. DAVEY and J. I. WILLIAMS	
39	*The Control of Technology*	D. ELLIOTT and R. ELLIOTT
40	*Cosmic Rays*	J. G. WILSON and G. E. PERRY
41	*Global Geology*	M. A. KHAN and B. MATTHEWS
42	†*Running, Walking and Jumping: The science of locomotion* A. I. DAGG and A. JAMES	
43	†*Geology of the Moon*	J. E. GUEST, R. GREELEY and E. HAY
44	† *The Mass Spectrometer*	J. R. MAJER and M. P. BERRY
45	† *The Structure of Planets*	G. H. A. COLE and W. G. WATTON
46	†*Images*	C. A. TAYLOR and G. E. FOXCROFT
47	† *The Covalent Bond*	H. S. PICKERING
48	†*Science with Pocket Calculators*	D. GREEN and J. LEWIS
49	†*Galaxies: Structure and Evolution*	R. J. TAYLER and A. S. EVEREST
50	†*Radiochemistry—Theory and Experiment*	T. A. H. PEACOCKE
51	†*Science of Navigation*	E. W. ANDERSON
52	†*Radioactivity in its historical and social context*	E. N. JENKINS and I. LEWIS
53	†*Man-made Disasters*	B. A. TURNER

†(*Paper and Cloth Editions available.*)

THE WYKEHAM ENGINEERING AND TECHNOLOGY SERIES

1   *Frequency Conversion*   J. THOMSON, W. E. TURK and M. J. BEESLEY
2   *Electrical Measuring Instruments*   E. HANDSCOMBE
3   *Industrial Radiology Techniques*   R. HALMSHAW
4   *Understanding and Measuring Vibrations*   R. H. WALLACE
5   *Introduction to Tribology*   J. HALLING and W. E. W. SMITH